NITRIDES

Sławomir Podsiadło

NITRIDES

Cover design and title pages
Przemysław Spiechowski

Publisher
Magdalena Ścibor

Editor
Avalon Languages
Maria Sala-Racinowska

Production coordinator
Mariola Iwona Keppel

The book was sponsored by the Faculty of Chemistry of the Warsaw University
of Technology.

ISBN 978-83-01-18166-6
Wydanie 1 – 1 dodruk

Polish Scientific Publishers PWN
G. Daimlera 2
02-676 Warsaw
Poland
tel. +48 22 69 54 321; fax +48 22 69 54 031
e-mail: pwn@pwn.com.pl; www.pwn.pl

For Beata, Jan, Ola and Greg

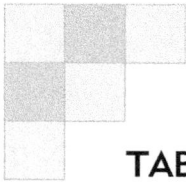

TABLE OF CONTENT

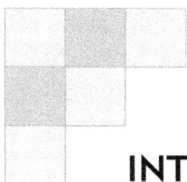

INTRODUCTION

Inorganic chemistry is a domain with probably the longest persisting traditional approach to the description of chemical compounds. In this traditional approach the highest importance is given to oxygen compounds, hence the compounds either existing in the lithosphere or those formed in the natural processes on Earth, even if they do not contain oxygen in their composition.

Recent decades brought a rapid development in research of new materials, especially in the field of spintronics or photovoltaics. Most often these are substances not occurring under natural conditions and they can be obtained only with the use of completely new preparation methods. Among the enormous number of currently known and used chemical compounds, one can distinguish certain groups of compounds of particular importance from the viewpoint of inorganic chemistry. These are, first of all, binary compounds, such as fluorides, sulfides, tellurides, nitrides, arsenides, phosphides, antimonides, silicides, and also complex ternary compounds – products of reactions of the binary compounds mentioned, such as fluoro- or thio-salts or complex nitrides. The latter are capable of forming many compounds with various elements of the periodic system, e.g. boron M_3BN_2, M_6BN_3, aluminum M_3AlN_2, gallium M_3GaN_2, silicon MSi_2N_3, M_2SiN_2, M_5SiN_3, M_8SiN_4, vanadium M_7VN_4 etc. (where M is an element of 1st group of the periodic system). One cannot disregard that among such compounds, not only containing nitrogen, one must look for still unresolved problems regarding the character of the relationship between the nature of chemical bonds or their relationship with the structure of microsystems. A weak side of chemistry may also result from its descriptive nature or, more precisely, the lack of a general theory enabling a uniform presentation of structures, properties and preparation methods of the enormous amount of compounds obtained hitherto. Such a state was also a limiting factor in theoretical forecasting. As late as with the development of the morphological classification, used in this monograph, it has become possible, at least in part, to go from an encyclopedic review of chemical compounds to a systematic treatment of chemical problems.

Nitrides have become in the recent decades probably the most fashionable group of simple chemical compounds, mainly owing to several compounds, such as boron nitride, gallium nitride and silicon nitride, all referred to as compounds of technological future. The cubic modification of BN is one the hardest chemical substances. As a semiconductor, GaN can replace silicon in many branches of traditional electronics and may serve as a basis for the developing spintronics. Si_3N_4 successfully replaces metal alloys in the construction of aircraft or motor-car engines, and replaces hard carbides in the edges of cutting tools.

This monograph should therefore play a binary role. First, it should present to the reader the possibility of transferring the relationships known in the domain of oxygen compounds to other groups of new chemical compounds, such as nitrides, and also to facilitate the preview of some phenomena and relationships, where the use of morphological classification of simple chemical species is very useful. On the other side, the monograph presents a wide range of preparation methods and basic chemical properties of known nitrides that can be significant from a practical viewpoint. However, the book will not be able to completely fill the large gap existing in the literature on the subject, since many nitrides have no sufficient reliable data regarding their structure that could be confirmed in the results of new studies.

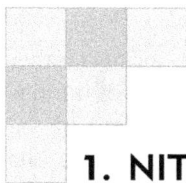

1. NITROGEN COMPOUNDS

1.1. Nitrides

The nitrides of the majority of elements have been known for a long time. However, since they did not find an immediate practical application, their physical and chemical properties were not precisely studied. Only in the last three decades have nitrides been studied intensively, when the electronic and space industries started to seek new materials. Boron nitride, the hexagonal, graphite-like form, is used as the source of boron for diffusion in silicon; the symmetrical, diamond-like nitride exhibits hardness comparable to that of diamond, and thus it is used as an abrasive. Aluminum and silicon nitrides are basic components of new ceramic materials of the SiAlON type, which are characterized by very high mechanical and thermal resistance. Gallium nitride has a possibility in the not too distant future of replacing silicon as a semiconductor, and silicon nitride the metal alloys as a construction material for appliances operating at high temperatures and in strongly corrosive media. Many nitrides possess specific properties, which are exploited in various fields of mechanics (e.g. Li_3N in cells with solid electrolyte, or P_3N_5 in vacuum technics). However, the nitrides of transition elements, of a characteristic structure (nitrogen is placed in the interstitial positions of the metal lattice), generally exhibit a considerable hardness and hence the common application of them in hardening the surface of machine parts. In monographs concerning nitrides they are divided into four groups, mainly on the basis of the kind of bond formed in the compound [1-5]. These are:

1. Nitrides of an ionic bond or an excess of ionic bond – these are formed by alkali metals, magnesium and alkaline earth metals, and elements of the copper and zinc subgroups.

2. Nitrides of covalent bonds or an excess of covalent bond – compounds of the sp elements.

3. Nitrides of metallic – covalent bond, where nitrogen is placed in the interstitial positions of the metal lattice – compounds of dsp and fdsp elements.

4. So-called mixed nitrides: reaction products of ionic nitrides with covalent ones. Actually, these are salts composed of a cation of an alkali metal or magnesium and alkaline earth metal and an anion in which, around the coordination center, a certain number of nitride ligands is placed-compounds of sp, dsp or fdsp elements.

This division does not seem to be consistent, and hence an attempt at another one, described later on, will be made. The order of discussing the properties of nitrides should be connected with the position in the periodic system of the element of the coordination center. It would be advantageous to simultaneously present all the nitrides of a given element, both at various oxidation states as well as of a different value of the nitrogen coordination number around the coordination center. This way of presentation differs from the one applied in monographs published until now, where the division into the four groups of nitrides mentioned earlier was quite strictly adhered to. As a result of this, boron nitride was described in a quite different section than Li_3BN_2 or Li_6BN_3 (and thus the direct products of the reaction of boron nitride with lithium nitride); and the latter ones were presented together with Li_9CrN_5 or Li_4VN_3. The application of a division other than a traditional one will permit this inconsistency. Thus, similarly as occurs in the typical description of oxy inorganic compounds, information on all the nitrides formed by a given element (also on complex compounds of the type of salts) can be found in the section devoted to that element.

The way of presenting compounds of successive elements is a separate problem. It is possible to apply the classical method, consisting in the description of the properties of consecutive compounds. The application of the morphological classification of simple species seemed to be of interest here [6, 7]. In the version proposed here it is possible to describe a molecule, or more exactly a species (which is a wider notion) of a chemical compound by means of two numbers precisely connected with the electronic and ligand structures surrounding the coordination center. These numbers, denoted by the symbols e_v and e_z, can be calculated from the following relationships:

$$e_v = Q_r - G_{ox} \qquad e_z = 2n$$

where: Q_r – charge of atomic core of coordination center
G_{ox} – oxidation state of coordination center
n – number of ligands

The e_v number denotes the number of electrons in the physical substance on the valence levels of the coordination center which do not participate in the formation of bonds with ligands. The e_z number denotes the number of electrons formally introduced by ligands to the coordination center and used for the formation of δ bonds. Multiple bonds of the π type are omitted in these considerations. In the majority, species introducing two electrons for bonding with the coordination center are ligands in coordination compounds. Due to the application of the numerical description of species of chemical compounds it is possible to place them on the classification table within the rectangular coordinate system $e_v - e_z$. The possibilities of the morphological classification for the ordering of structures and description of their transformations – both in the substance of the methods of their preparation and reactivity – will be presented using the example of silicon (an element of atomic core charge 4+, which is equal to the number of valence electrons in the electroneutral atom). Silicon forms a relatively large number of nitride species and thus it is a convenient subject for the presentation of general relationships.

In Fig. 1. the classification table in the $e_v - e_z$ coordinate system with the monocentric (having one coordination center) nitride species of silicon is presented. The morphological classification, in addition to permitting the ordering of the presented structures, is arranged in such a way that the transformations joining the particular species are of the character of elementary chemical processes. All processes proceeding among the species placed in the table will be presented in the section devoted to nitride species of silicon. In this section, however, only chosen elementary transformations are presented.

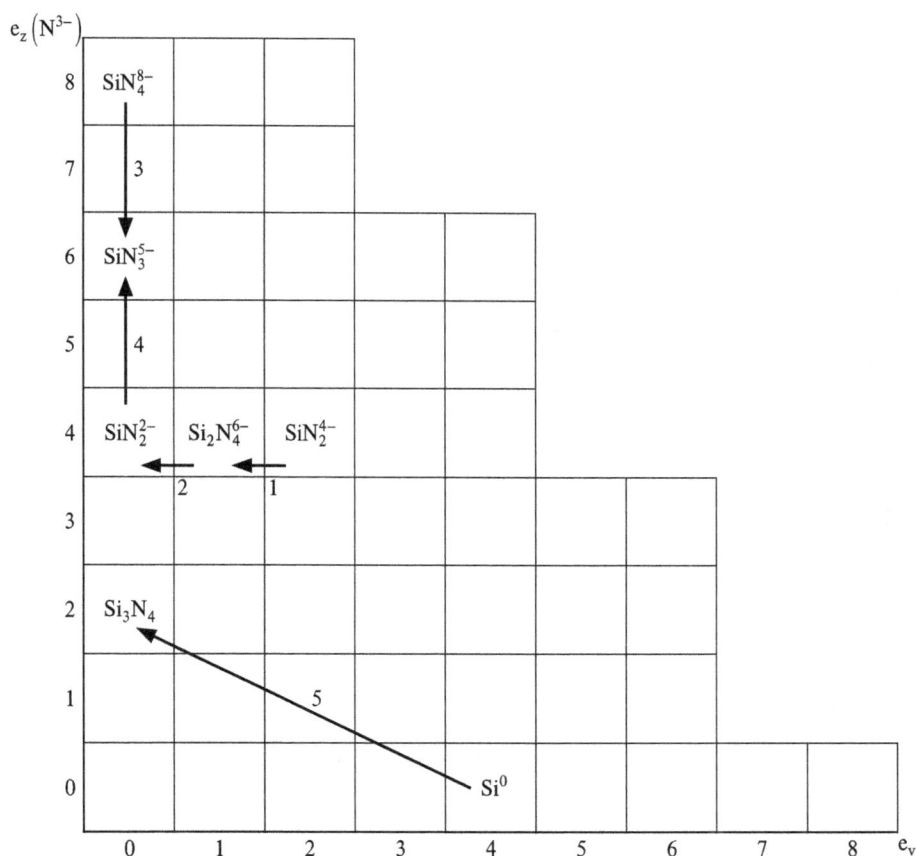

Fig. 1. Nitride compounds of silicon

The thermal decomposition of H_4SiN_2 leads to H_2SiN_2 with hydrogen evolution. When it is presented by means of protonless [6, 7] skeletons the process can be formally treated as a pure deelectronization one:

1. $2SiN_2^{4-} \rightarrow Si_2N_4^{6-} + 2e$

2. $Si_2N_4^{6-} \rightarrow 2SiN_2^{2-} + 2e$

i. e. a redox transformation.

On the other hand, the condensation of silicon tetraamide with ammonia evolution can be treated as an acid–basic process, in this case deanionization:

3. $SiN_4^{8-} \rightarrow SiN_3^{5-} + N^{3-}$

The reaction of the mixed lithium–silicon nitride with lithium nitride, consisting in addition of the N^{3-} anion, is a reverse ac-bas process:

4. $SiN_2^{2-} + N^{3-} \rightarrow SiN_3^{5-}$

Coupled processes, in the morphological classification substance, should be understood as a simultaneous acidic-basic and oxidative-reductive acts. The synthesis of silicon nitride from free elements is an example of such a process, where silicon is an electron donor (red) and acceptor of the nitride anion formed (ac), and nitrogen is an electron acceptor (ox) and donor of N^{3-} anions (bas):

5. $3Si + 2N_2 \rightarrow Si_3N_4$

Thus, due to the application of the morphological classification of simple species, it becomes possible to synthetically present the structures, methods of preparation and properties of nitrides of successive elements.

In this book these species, their properties and methods of preparation are presented in the following order:

1. ionic nitrides in groups of the Mendeleyev system
2. covalent nitrides in the order of increasing electronegativity in the periods of the Mendeleyev system (including complex nitrides)
3. nitrides of transition and intratransition elements (including complex nitrides).

The description of ionic as well as covalent species should also cover amides and imides, i.e. compounds in which (in comparison with nitride) part of the cations (or cationides) is replaced by protons. Such a formulation will approximate the way of presentation to the classical one, applied in inorganic chemistry for oxy species, where also hydroxides, acids or hydrogen salts are presented besides oxides or salts. Azides – peroxides analogs among oxy compounds will be described for a complete presentation of inorganic compounds in which nitrogen is the ligand (anionide). However, comparatively less attention will be given to those compounds.

1.2. Oxynitride compounds

Some doubts may arise from the presentation of compounds with nitrogen as the ligand. Compounds are known, however, in which a mixed oxynitride coordination surrounding round the central element occurs. Compounds of this type (called oxynitrides) are formed by aluminum (AlON phases), silicon (Si_2N_2O), germanium (Ge_2N_2O) or phosphorus (PNO). Salts containing anions of a mixed ligand coordination center are also known: cyanates (CNO^-) and the oxynitridesilicates ($SiNO^-$) or oxynitridegermanates ($GeNO^-$).

Thus, compounds containing nitride ligands in the surrounding of the coordination center together with other ligands, not only oxide ones, will be presented in a limited scale. For the presentation of compounds of a given element of a mixed coordination center it is necessary to apply the modified version of the morphological classification. In comparison with the classification system of purely nitride compounds (see section 1.1) the introduction of the third axis leads to a three-dimensional space in the coordinate system $e_v - e_z(O^{2-}) - e_z(N^{3-})$ where the respective numbers denote: e_v – number of valence electrons at the coordination center, $e_z(O^{2-})$ – number of z electrons formally introduced to the coordination center and used for the formation of δ bonds by oxide ligands, $e_z(N^{3-})$ – number of electrons formally introduced to the coordination center and used for the formation of δ bonds by nitride ligands. If assuming that no changes occur in the oxidation state of element A during transformations of its oxynitride compound, the classification system can be simplified to a two-dimensional one of the $e_z(O^{2-}) - e_z(N^{3-})$ axes.

In Fig. 2. the classification table of oxynitride anions of the hypothetic element A at the n+ oxidation state (charge of core A^{n+}) are presented. Transformations are also presented, in which the ANO^{n-5} anion can participate. The following reactions can take place:

1. $ANO^{n-5} + N^{3-} \rightarrow AN_2O^{n-8}$

2. $ANO^{n-5} + O^{2-} \rightarrow ANO_2^{n-7}$

3. $ANO^{n-5} \rightarrow AN^{n-3} + O^{2-}$

4. $ANO^{n-5} \rightarrow AO^{n-2} + N^{3-}$

5. $2ANO^{n-5} \rightarrow AN_2^{n-6} + AO_2^{n-4}$

6. $AN_2^{n-6} + AO_2^{n-4} \rightarrow 2ANO^{n-5}$

The numbers of the reactions correspond to those of the transformations in the table in Fig. 2.

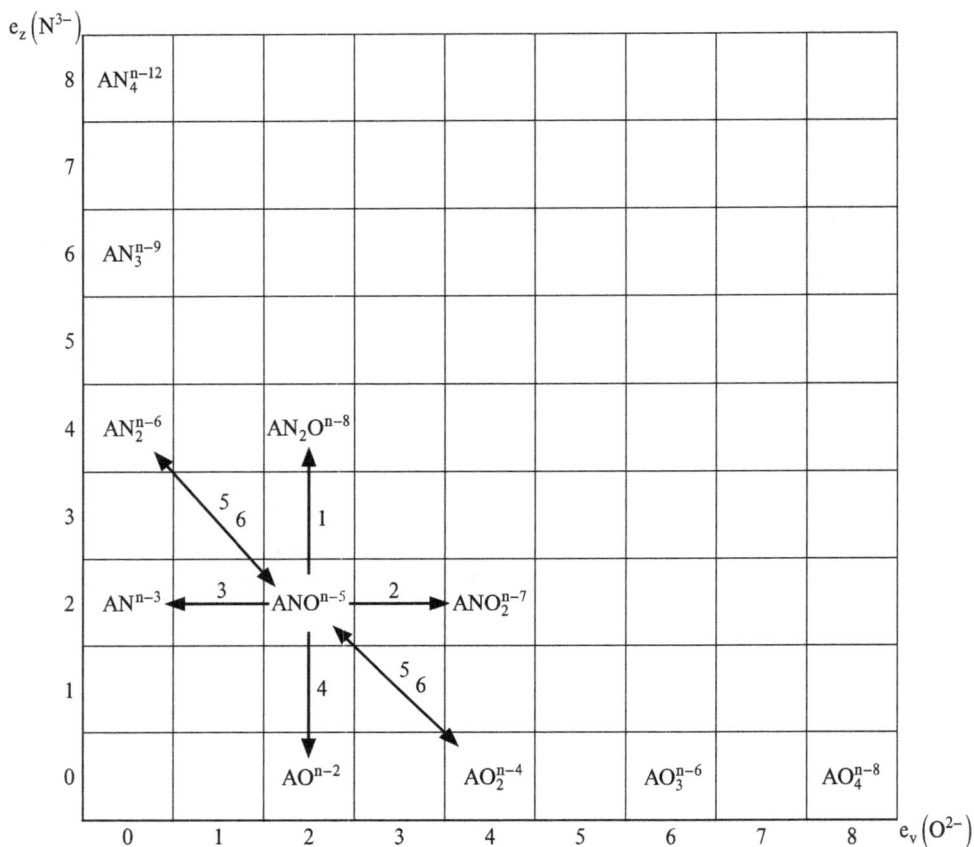

Fig. 2. Oxynitride anions of hypothetic element A

Reactions 1 and 2 can be defined as anionization processes (addition of a respective anion), and reactions 3 and 4 as deanionization ones. Transformations 5 and 6 are called acidic-basic disproportionation and synproportionation, respectively. In such a presented classification system (both axes based on e_z numbers) it is possible to describe exclusively acidic-basic transformations, understood quite widely, based on the Gutman-Lindquist ionotropic theory. The six types of reactions presented above are sufficient for a general presentation of all transformations possible within the range of the described hypothetical oxynitride compounds of element A. All the more complex processes are a combination of the ones mentioned above.

The group of oxynitride compounds of an element is a set of compounds limited by the maximum coordination number, in the region of which some theoretical predictions of transformations are possible. As can be easily noticed, when considering monocentric (having one center) species, then sites of the values of the $e_z(O^{2-})$ number are equal to 2n and the $e_z(N^{3-})$ number also are equal to 2n, where n = 1,2,3,4 are accessible for them. It is possible to locate in the table all the hypothetic anions of a respective element and to present on the table the routes of their preparation. The verification of the hypotheses of the existence of new chemical compounds is made by means of experimental studies. The oxynitride compounds known until now will be briefly described when discussing the compounds of the successive elements.

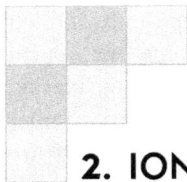

2. IONIC NITRIDES

The low electronegativity of alkali metals and the strong electron donor properties connected with it cause that in reactions with nitrogen these metals form ionic type compounds of the general formula $M_3^I N$. The formation of other compounds is possible, in which an electron deficiency occurs at nitrogen. The anionic part of the complex then adopts a more complicated structure of the 16 electron N_3^- species. The stability of azides of alkali metals increased with an increase of the atomic number, and that of nitrides decreased. Despite that the lowering of electronegativity indicates a reverse direction of changes. This effect is explained by the different electronic structure of potassium and consecutive elements, in which the d energetic levels are completely filled. This, considering their high symmetry, decreases the energy of the whole system and, as a consequence, permits the back bonding of the N_3^- anion (delocalization of the p electron of nitrogen). On the other hand, however, the effect of the cation on the azide structure is minimal, which is exhibited by a nearly identical pattern of IR spectra of the majority of azides [1, 2]. All elements of the first and second group of the periodic table form azides. In so far as hydrazoic acid has a non linear, asymmetric structure the N_3^- anion in all the azides has a linear and symmetric one. In so far as the alkali metals azides are stable, the azides of beryllium, magnesium and alkaline earth metals as well as of heavy metals explode at elevated temperature or when struck. Alkali metal nitrides are of a hexagonal or tetragonal crystalline structure.

A similar type of compounds is formed with nitrogen by copper, silver and gold. The electronegativity of these elements is indeed much greater than that of alkali metals, but the symmetric structure of the electronic shell in M^+ cations causes an ionic bond to occur also in these compounds. The copper family nitrides, however, are not dielectrics any more, but, contrary to alkali metal nitrides, they exhibit semi-conducting properties. They all crystallize in a regular system.

Beryllium, magnesium and alkaline earth metals form nitrides of the $M_3^{II} N_2$ composition of an ionic bonding, being dielectrics. Starting from calcium, compounds of another composition are known, e.g. Ca_2N, Ca_3N_4, etc.

The assignment of potassium, rubidium and cesium from among the alkali metals and of calcium, strontium and barium from among beryllium, magnesium and alkaline earth metals to the sp, dsp or fdsp block of elements is of minor importance in relation to the problems described here, but interesting from the point of view of general chemistry.

The kind of nitrides formed by alkali metals – all of a $M_3^I N$ stoichiometry, does not warrant introducing differentiations among elements forming cations. Strontium and barium, however, besides nitrides typical for their group, form $M_2^{II} N$ compounds known for quite a long time, of a metallic character of bonds and interstitial structure. Calcium nitride of the Ca_2N composition was obtained relatively recently. Compounds of this type are formed by a majority of transition elements, which will be described in one of the successive sections. Zinc, cadmium and mercury nitrides are of the M_3N_2 composition and have a regular structure. Metal-nitrogen bonds are of an ionic character, and the compounds exhibit semi-conducting properties. All the elements described in this section form amides and imides. Attention should be drawn to analogies occurring between hydroxide and azoimide compounds of alkali metals and beryllium, magnesium and alkaline earth metals:

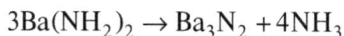

$$3Ba(NH_2)_2 \rightarrow Ba_3N_2 + 4NH_3$$

$$3BaNH \rightarrow Ba_3N_2 + NH_3$$

It should be noted that all the reactions of nitrides preparation and also of their transformations should be carried out under an oxygen-free atmosphere (oxygen, being more electronegative than nitrogen, is stronger bonded in compounds).

2.1. Nitrides of alkali metals

2.1.1. Lithium nitrides

Lithium, according to the present state of knowledge, forms only one nitride of the formula Li_3N. It has a hexagonal structure which is showed in Fig. 3.

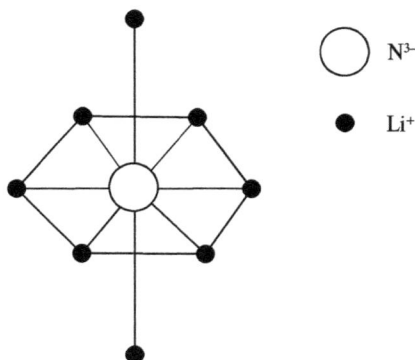

Fig. 3. The structure of lithium nitride

Lithium nitride is a chemically unstable compound. At room temperature it undergoes hydrolysis transforming into hydroxide with ammonia evolution. However, Li_3N is a strong nitride anion donor in reactions with nitrides of elements characterized by moderate electronegativity. Complex nitrides are formed in these processes – compounds, in which lithium cations form salts with anions of a nitride environment round the coordination center. The central element can be represented both by the main family elements and by transition and intratransition ones, which will be described in chapters devoted to successive elements. Lithium nitride reacts at an elevated temperature with hydrogen forming lithium hydride:

$$Li_3N + 3H_2 \rightarrow 3LiH + NH_3$$

and with nitrogen oxide forming a mixture of solid products: Li_2O, Li_2O_2, $LiNO_2$ and gaseous N_2. At 400°C in the presence of mercury, Li_3N undergoes decomposition with nitrogen evolution and formation of the LiHg amalgam. Lithium nitride is obtained from the reaction of metallic lithium with nitrogen at a temperature of up to 450°C. Above that temperature a strong exothermic reaction proceeds in an uncontrollable manner. 170°C was found to be the optimum temperature, i.e. the one nearest to the melting point of metallic lithium. On the other hand, an increase of nitrogen pressure to 6 atmospheres makes the process more effective. The reaction of metallic lithium with ammonia leads to the amide and next to imide:

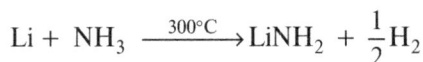

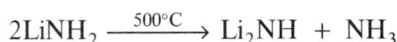

$$Li + NH_3 \xrightarrow{300°C} LiNH_2 + \frac{1}{2}H_2$$

$$2LiNH_2 \xrightarrow{500°C} Li_2NH + NH_3$$

The presence of lithium nitride was found in this process up to 900°C.
Lithium azide LiN_3 undergoes decomposition at above 200°C.

$$3LiN_3 \rightarrow Li_3N + 4N_2$$

The azide LiN_3 is obtained by precipitation from a saturated solution of lithium hydroxide with a solution of hydrazoic acid, HN_3.

2.1.2. Sodium nitrides

Sodium nitride, Na_3N, is a not very stable compound. It undergoes thermal decomposition to elements at 270°C (in an inert atmosphere). Heated in oxygen it transforms into sodium oxide with nitrogen evolution:

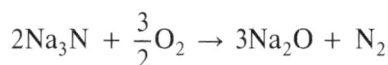

$$2Na_3N + \frac{3}{2}O_2 \rightarrow 3Na_2O + N_2$$

22

It vigorously hydrolyzes at normal temperature with ammonia evolution:

$$Na_3N + 3H_2O \rightarrow 3NaOH + NH_3$$

Sodium nitride also undergoes reaction with chlorine, sulfur and phosphorus:

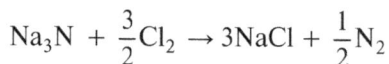

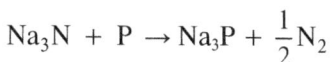

$$Na_3N + \frac{3}{2}Cl_2 \rightarrow 3NaCl + \frac{1}{2}N_2$$

$$2Na_3N + 3S \rightarrow 3Na_2S + N_2$$

$$Na_3N + P \rightarrow Na_3P + \frac{1}{2}N_2$$

Na_3N is obtained during electric discharges in a closed system containing sodium vapors in a nitrogen atmosphere under reduced pressure.

The reaction of sodium hydride with nitrogen at 200°C also leads to the nitride:

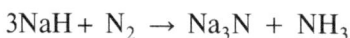

$$3NaH + N_2 \rightarrow Na_3N + NH_3$$

The reaction of metallic sodium with ammonia results not in the nitride but the amide:

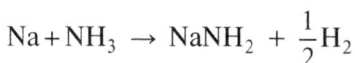

$$Na + NH_3 \rightarrow NaNH_2 + \frac{1}{2}H_2$$

Sodium azide is applied in the synthesis of the majority of the other azides. It does not hydrolyze in water at normal temperature. Water vapor causes its decomposition at a high temperature according to the reaction:

$$3NaN_3 + 3H_2O \rightarrow 3NaOH + NH_3 + 4N_2$$

NaN_3 can be obtained from the following processes:

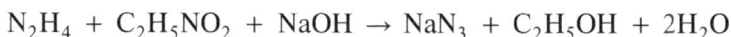

$$2NaNH_2 + N_2O \rightarrow NaN_3 + NaOH + NH_3$$

$$NaNO_3 + 3NaNH_2 \rightarrow NaN_3 + 3NaOH + NH_3$$

$$N_2H_4 + C_2H_5NO_2 + NaOH \rightarrow NaN_3 + C_2H_5OH + 2H_2O$$

$$Na + NH_4N_3 \rightarrow NaN_3 + NH_3 + \frac{1}{2}H_2$$

2.1.3. Potassium nitrides

Potassium nitride K_3N is a compound of scarcely studied properties. It is very difficult to obtain it in the pure state. During heating in an inert atmosphere it undergoes decomposition to elements. Similarly as sodium nitride, it undergoes reaction with chlorine, oxygen, sulfur and phosphorus with nitrogen evolution and formation of the chloride, oxide, sulfide or phosphide, respectively. It rapidly hydrolyzes in water transforming into potassium hydroxide with ammonia evolution. Potassium nitride can be obtained similarly as sodium nitride, from the reaction of atomic nitrogen, formed during electric discharges, with potassium vapors under reduced pressure. However, the mixture of nitride and azide is obtained under these conditions. K_3N is formed during nitrogen flow over potassium hydride:

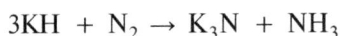

$$3KH + N_2 \rightarrow K_3N + NH_3$$

and also during the thermal decomposition of the azide under vacuum:

$$3KN_3 \rightarrow K_3N + 4N_2$$

However, no pure product of a composition corresponding to the theoretical one is obtained in either of these reactions. Metallic potassium in reaction with ammonia forms an amide which transforms in elevated temperature into the imide:

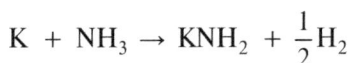

$$K + NH_3 \rightarrow KNH_2 + \frac{1}{2}H_2$$

$$2KNH_2 \rightarrow K_2NH + NH_3$$

Potassium azide is well soluble in water, however, at an elevated temperature it slowly hydrolyzes with nitrogen and ammonia evaluation. KN_3 melts not sharply with simultaneous decomposition. Potassium azide can be obtained similarly as sodium salt, i.e. in the reaction of amide with nitrogen suboxide, potassium nitrate with potassium amide, hydrazine with nitroethane and potassium hydroxide, and ammonium azide with metallic potassium (the reactions for sodium compounds are presented in the previous section).

2.1.4. Rubidium nitrides

Rubidium nitride is a chemically more stable compound than sodium or potassium nitrides. In water it undergoes slow hydrolysis:

$$Rb_3N + 3H_2O \rightarrow 3RbOH + NH_3$$

It does not react at room temperature with oxygen, chlorine, sulfur or phosphorus. Reactions with these elements take place at above 100°C, e.g.:

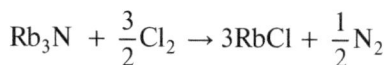

$$Rb_3N + \frac{3}{2}Cl_2 \rightarrow 3RbCl + \frac{1}{2}N_2$$

In a hydrogen atmosphere the nitride transforms into a hydride:

$$Rb_3N + 3H_2 \rightarrow 3RbH + NH_3$$

Rubidium nitride is obtained from the reaction of rubidium hydride and nitrogen at above 300°C:

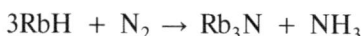

$$3RbH + N_2 \rightarrow Rb_3N + NH_3$$

The decomposition of rubidium azide at 340°C also leads to a nitride:

$$3RbN_3 \rightarrow Rb_3N + 4N_2$$

Rubidium azide is well soluble in water, in which it does not hydrolyze. It decomposes at an elevated temperature under normal pressure, as mentioned above, to a nitride, and in vacuum conditions to metal:

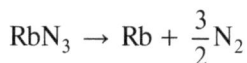

$$RbN_3 \rightarrow Rb + \frac{3}{2}N_2$$

Rubidium azide is formed in the reaction of solid rubidium carbonate or solid rubidium hydroxide with ammonia. It can also be obtained in an aqueous solution by precipitating barium sulfate after mixing the solutions of Rb_2SO_4 and $Ba(N_3)_2$.

2.1.5. Cesium nitrides

Cesium nitride reacts violently with water with ammonia evolution. In an anhydrous atmosphere it is stable at room temperature in hydrogen, oxygen and chlorine. At higher temperatures hydrogen causes the transformation of nitride into a hydride:

$$Cs_3N + 3H_2 \rightarrow 3CsH + NH_3$$

Cesium nitride is formed from the reaction of nitrogen with cesium hydride at 340°C (the reaction proceeds similarly as in the case of the alkali metal compounds described previously). Cesium azide is a stable compound at normal temperature in the air and in aqueous solutions. It transforms in a hydrogen atmosphere at elevated temperatures into a hydride, and at 340°C in nitrogen it decomposes to the nitride Cs_3N. At the same temperature, under vacuum conditions, it decomposes to pure metal.

2.2. Beryllium, magnesium and alkaline earth metal nitrides

2.2.1. Beryllium nitrides

Beryllium forms only one nitride of the Be_3N_2 composition. In the crystalline state it has a regular structure of the anti-Mn_2O_3 type of the lattice constant a = 8.15Å. Beryllium nitride melts at 2200°C with simultaneous slow decomposition. In an oxygen and chlorine atmosphere it is stable up to ca. 500°C, and in an atmosphere of gaseous hydrochloride – up to 700°C. The following reactions proceed above these temperatures:

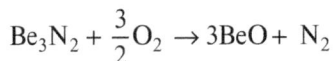

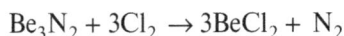

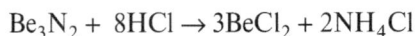

$$Be_3N_2 + \frac{3}{2}O_2 \rightarrow 3BeO + N_2$$

$$Be_3N_2 + 3Cl_2 \rightarrow 3BeCl_2 + N_2$$

$$Be_3N_2 + 8HCl \rightarrow 3BeCl_2 + 2NH_4Cl$$

Fused salts react with beryllium nitride above 1200°C. Hydrogen does not react up to 1000°C. Beryllium nitride is obtained in the reaction of free beryllium with nitrogen at 500–900°C. The process proceeds very slowly due to the formation on the surface of beryllium of a compact nitride layer limiting the diffusion of nitrogen. Similar problems occur when using ammonia as the nitriding agent; the process is then carried out at 700–1000°C:

$$3Be + N_2 \rightarrow Be_3N_2$$

$$3Be + 2NH_3 \rightarrow Be_3N_2 + 3H_2$$

Beryllium nitride, contaminated with carbon, beryllium oxide and carbide, can be obtained by reducing beryllium oxide with carbon with simultaneous nitriding in a nitrogen flux:

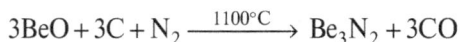

$$3BeO + 3C + N_2 \xrightarrow{1100°C} Be_3N_2 + 3CO$$

Beryllium carbide transforms into a nitride under the action of nitrogen or ammonia at above 1250°C:

$$3Be_2C + 2N_2 \rightarrow 2Be_3N_2 + 3C$$

Beryllium azide, $Be(N_3)_2$, decomposes to a nitride at elevated temperatures with nitrogen evolution; on the contrary to the magnesium and alkali earth metal azides –

without an explosion. It easily decomposes under the influence of water. The azide is obtained in the reaction of dimethylberyllium with hydrazoic acid at –150°C:

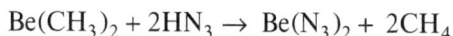

$$Be(CH_3)_2 + 2HN_3 \rightarrow Be(N_3)_2 + 2CH_4$$

2.2.2. Magnesium nitrides

Magnesium forms only one nitride of Mg_3N_2 composition crystallizing in the regular form of the lattice constant a = 9.95Å. Mg_3N_2 melts at 819°C without decomposition. It undergoes violent oxidation in the atmosphere of oxygen above 400°C:

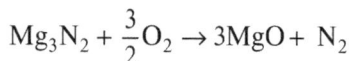

$$Mg_3N_2 + \frac{3}{2}O_2 \rightarrow 3MgO + N_2$$

Chlorine reacts with magnesium nitride at 300°C:

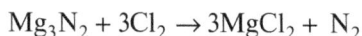

$$Mg_3N_2 + 3Cl_2 \rightarrow 3MgCl_2 + N_2$$

Mg_3N_2 undergoes hydrolysis in water with ammonia evolution:

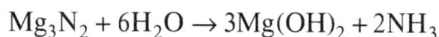

$$Mg_3N_2 + 6H_2O \rightarrow 3Mg(OH)_2 + 2NH_3$$

Magnesium nitride is a reactive compound – strong donor of nitride anions, both in the reaction of phosphorus pentachloride:

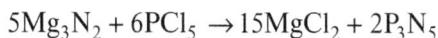

$$5Mg_3N_2 + 6PCl_5 \rightarrow 15MgCl_2 + 2P_3N_5$$

and in reaction with nitrides of elements of moderate electronegativity values, forming complex nitrides with them, e.g.:

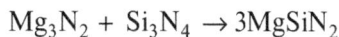

$$Mg_3N_2 + Si_3N_4 \rightarrow 3MgSiN_2$$

Mg_3N_2 exhibits strong reducing properties:

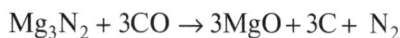

$$Mg_3N_2 + 3CO \rightarrow 3MgO + 3C + N_2$$

$$2Mg_3N_2 + 3CO_2 \rightarrow 6MgO + 3C + 2N_2$$

Magnesium nitride can be obtained in the reaction of magnesium with gaseous nitrogen at above 500°C:

$$3Mg + N_2 \rightarrow Mg_3N_2$$

Magnesium reacts with ammonia under similar conditions:

$$3Mg + 2NH_3 \rightarrow Mg_3N_2 + 3H_2$$

Magnesium hydride in reaction with nitrogen forms a nitride:

$$3MgH_2 + 2N_2 \rightarrow Mg_3N_2 + 2NH_3$$

The reduction of cyanides with magnesium above 1000°C yields magnesium nitride and carbon in the post-reaction mixture:

$$2KCN + 3Mg \rightarrow Mg_3N_2 + 2K + 2C$$

Nowadays, only the synthesis of the nitride from metallic magnesium is of practical importance.

2.2.3. Calcium nitrides

Calcium forms a number of compounds with nitrogen of different compositions: Ca_2N, Ca_3N_2, Ca_3N_4 and $Ca(N_3)_2$.

The nitride Ca_2N is probably one of the many interstitial compounds among the transition elements, which were mentioned in the introduction to this chapter.

A metallic bond occurs, which is reflected in the course of reactions with water or acids, in which hydrogen evolves:

$$Ca_2N + 4H_2O \rightarrow 2Ca(OH)_2 + NH_3 + \frac{1}{2}H_2$$

Ca_2N is formed from the decomposition of Ca_3N_2 at 1100°C under reduced pressure:

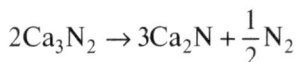

$$2Ca_3N_2 \rightarrow 3Ca_2N + \frac{1}{2}N_2$$

Ca_3N_2 is formed during direct nitriding with nitrogen of metallic calcium at 350–1050°C. Different crystalline forms of the nitride are formed depending on the temperature: at 350°C – β-Ca_3N_2 (hexagonal form), at 700°C – α-Ca_3N_2 (regular form) and at about 1050°C – γ-Ca_3N_2 (rhombic form). Ca_3N_2 melts under normal pressure at 1200°C, and from 1270°C it slowly decomposes. Calcium nitride, Ca_3N_2, is obtained exclusively from metallic calcium in reaction with nitrogen or ammonia:

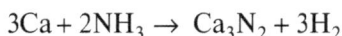

$$3Ca + N_2 \rightarrow Ca_3N_2$$

$$3Ca + 2NH_3 \rightarrow Ca_3N_2 + 3H_2$$

The nitride of the Ca_3N_4 composition and of not precisely determined properties is formed from the decomposition of calcium amide during heating under vacuum:

$$3Ca(NH_2)_2 \rightarrow Ca_3N_4 + 2NH_3 + 3H_2$$

Calcium azide, $Ca(N_3)_2$, can be obtained only by precipitation from aqueos solutions of calcium salts with hydrazoic acid.

2.2.4. Strontium nitrides

Strontium, similarly as calcium, forms a number of compounds with nitrogen of different compositions: Sr_2N, SrN, Sr_3N_2, Sr_3N_4 and $Sr(N_3)_2$. Sr_2N and probably SrN are interstitial compounds, in which nitrogen is inbuilt in the interstitial positions of the metal crystalline lattice, but the metallic bonds between the strontium atoms still maintain their character. This is reflected in the reactions of the M_2N type nitrides with acids or water, where hydrogen evolves:

$$Sr_2N + 5HCl \rightarrow 2SrCl_2 + NH_4Cl + \frac{1}{2}H_2$$

Sr_3N_2 is stable in an anhydrous atmosphere, but in water it undergoes hydrolysis with the formation of strontium hydroxide and ammonia evolution, i.e. identically as in the case of magnesium and calcium nitrides of the same composition. From the reaction of metallic strontium with nitrogen or ammonia at above 400°C the nitride Sr_3N_2 is formed:

$$3Sr + N_2 \rightarrow Sr_3N_2$$

$$3Sr + 2NH_3 \rightarrow Sr_3N_2 + 3H_2$$

Sr_3N_2 decomposes to Sr_2N under vacuum at above 450°C:

$$2Sr_3N_2 \rightarrow 3Sr_2N + \frac{1}{2}N_2$$

Also the decomposition of the azide leads to the nitride:

$$3Sr(N_3)_2 \rightarrow Sr_3N_2 + 8N_2$$

Sr_3N_4 is obtained during the decomposition of strontium hexammine.

Alkali metals, calcium and alkaline earth metals dissolve in liquid ammonia forming solutions of a metallic gloss and high electric conduction. Ammines can be isolated by crystallization from these solutions, e.g.: $Li(NH_3)_4$ or $Sr(NH_3)_6$.

$$Sr(NH_3)_6 \rightarrow Sr(NH_2)_2 + 4NH_3 + H_2$$

$$3Sr(NH_2)_2 \rightarrow Sr_3N_4 + 2NH_3 + 3H_2$$

Strontium amide is an intermediate product of the decomposition, and besides Sr_3N_4 also an imide is obtained, which is formed in the reaction:

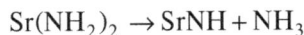

$$Sr(NH_2)_2 \rightarrow SrNH + NH_3$$

The azide is obtained by treating a strontium hydroxide solution with hydrazoic acid.

2.2.5. Barium nitrides

Barium forms the following compounds with nitrogen: Ba_2N, BaN, Ba_3N_2, Ba_3N_4, BaN_2 and $Ba(N_3)_2$, i.e. very similar to the strontium group of compounds, with only an extra nitride of BaN_2 composition. Both the properties and preparation methods of corresponding barium and strontium nitrides are very similar. Ba_2N has the properties of a compound with metallic bonds:

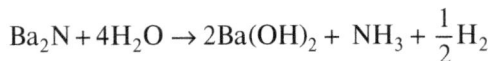

$$Ba_2N + 4H_2O \rightarrow 2Ba(OH)_2 + NH_3 + \frac{1}{2}H_2$$

Ba_3N_2 is stable at normal temperature in an anhydrous atmosphere. In water it undergoes hydrolysis:

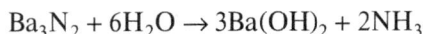

$$Ba_3N_2 + 6H_2O \rightarrow 3Ba(OH)_2 + 2NH_3$$

The properties of Ba_3N_4 and BaN_2 are poorly studied. Ba_3N_2 is obtained by nitriding metallic barium with nitrogen at 280–600°C or with ammonia at 400°C:

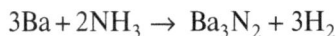

$$3Ba + N_2 \rightarrow Ba_3N_2$$

$$3Ba + 2NH_3 \rightarrow Ba_3N_2 + 3H_2$$

Ba_3N_2 undergoes decomposition under vacuum at 500°C:

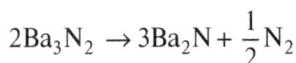

$$2Ba_3N_2 \rightarrow 3Ba_2N + \frac{1}{2}N_2$$

The decomposition of barium azide leads to Ba_3N_4:

$$3Ba(N_3)_2 \rightarrow Ba_3N_4 + 7N_2$$

On the other hand, BaN_2 is formed from Ba_3N_2 under high nitrogen pressure at 500°C:

$$Ba_3N_2 + 2N_2 \rightarrow 3BaN_2$$

2.3. Nitride compounds of the elements of the copper subgroup

2.3.1. Copper nitrides

Copper with nitrogen forms the nitride Cu_3N and azides CuN_3 and $Cu(N_3)_2$. Cu_3N has a regular crystalline structure of a lattice constant a = 3.807Å. It is not a very stable compound, slowly decomposing in water with ammonia evolution. It oxidizes in air above 400°C:

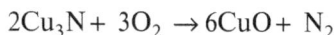

$$2Cu_3N + 3O_2 \rightarrow 6CuO + N_2$$

In a nitrogen atmosphere it undergoes decomposition to elements at above 500°C:

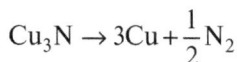

$$Cu_3N \rightarrow 3Cu + \frac{1}{2}N_2$$

None of the copper nitrides, silver and gold were obtained until now from the direct reaction of nitrogen with the metal. They are obtained by treating corresponding oxides with ammonia – both in the gas phase and with ammonia water. Cu_3N is formed in the following reactions with gaseous ammonia:

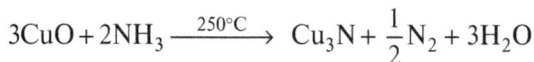

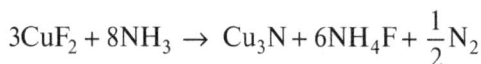

$$3CuO + 2NH_3 \xrightarrow{250°C} Cu_3N + \frac{1}{2}N_2 + 3H_2O$$

$$2Cu_2O + 2NH_3 \xrightarrow{250°C} 2Cu_3N + 3H_2O$$

$$3CuF_2 + 8NH_3 \rightarrow Cu_3N + 6NH_4F + \frac{1}{2}N_2$$

The azides CuN_3 and $Cu(N_3)_2$ are obtained by precipitation from an aqueous solution of corresponding salts. They are relatively stable – they do not exhibit explosive properties.

2.3.2. Silver nitrides

Two compounds of silver with nitrogen are known: nitride Ag_3N and azide AgN_3. Ag_3N is scarcely soluble in water and does not hydrolyze. It explodes very easily when struck or heated above 150°C. It crystallizes in a regular form – lattice constant $a = 4.378$Å.

Free nitrogen does not react with silver, Ag_3N can be obtained by very rapid cooling, from 1300°C to room temperature, of silver vapors in an ammonia atmosphere. The thermal decomposition of $AgF \cdot 2NH_3$ up to 100°C leads to the nitride:

$$3AgF \cdot 2NH_3 \rightarrow Ag_3N + 3NH_4F + 2NH_3$$

Ag_3N slowly crystallizes from a solution of concentrated ammonia saturated with silver oxide:

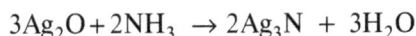

$$3Ag_2O + 2NH_3 \rightarrow 2Ag_3N + 3H_2O$$

Also the azide AgN_3 has explosive properties and it is precipitated from the aqueous solution of silver nitrate with sodium azide.

2.3.3. Gold nitrides

Two gold nitrides, Au_3N and Au_3N_2, have already been obtained. These compounds are unstable – at elevated temperature they decompose with explosion. Gold nitrides are obtained in reactions of corresponding oxides with ammonia:

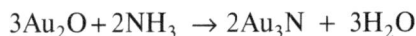

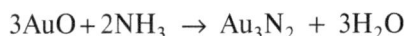

$$3Au_2O + 2NH_3 \rightarrow 2Au_3N + 3H_2O$$

$$3AuO + 2NH_3 \rightarrow Au_3N_2 + 3H_2O$$

2.4. Nitride compounds of the elements of the zinc subgroup

2.4.1. Zinc nitrides

Zinc forms two compounds with nitrogen: nitride Zn_3N_2 and azide $Zn(N_3)_2$. Zn_3N_2 in the crystalline form has a regular structure of the anti-Mn_2O_3 type of the lattice constant $a = 9.78$Å. It decomposes to free elements in nitrogen under normal pressure at above 600°C or under vacuum at above 300°C:

$$Zn_3N_2 \rightarrow 3Zn + N_2$$

Zinc nitride rapidly hydrolyzes in water with ammonia evolution. Zn_3N_2 is formed during the nitriding of metallic zinc with ammonia at temperatures from 550°C to 600°C:

$$3Zn + 2NH_3 \rightarrow Zn_3N_2 + 3H_2$$

The thermal decomposition of zinc amide leads to nitride with ammonia evolution at above 250°C. On the other hand, zinc cyanide in reaction with a large excess of ammonium nitrate at elevated temperature yields zinc nitride among the solid products, however contaminated with free zinc, oxide and cyanide. Zinc nitride can also be obtained by rapid cooling of a mixture of nitrogen and zinc vapors in a reactor at 1000°C, in which constant electric discharges are applied. Zinc azide is obtained from a water solution of zinc salt with hydrazoic acid or sodium azide. $Zn(N_3)_2$ has explosive properties.

2.4.2. Cadmium nitrides

Until now two compounds of cadmium with nitrogen have been obtained: nitride Cd_3N_2 and azide $Cd(N_3)_2$.

Cd_3N_2 has a regular structure of a lattice constant a = 10.79Å. In a nitrogen atmosphere at above 300°C it undergoes decomposition to elements. In the presence of oxygen it transforms at elevated temperature into cadmium oxide with nitrogen evolution. The decomposition of cadmium amide leads to a nitride, similarly as in the case of zinc amide:

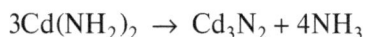

$$3Cd(NH_2)_2 \rightarrow Cd_3N_2 + 4NH_3$$

but it is advisable to carry out the process under vacuum at 180°C.

Cadmium oxide, when heated in ammonia flux at below 300°C, very slowly transforms into the nitride:

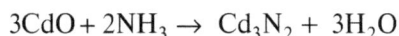

$$3CdO + 2NH_3 \rightarrow Cd_3N_2 + 3H_2O$$

Cadmium azide $Cd(N_3)_2$, similarly as zinc azide, is obtained by preparative methods from aqueous solutions. It is a strongly explosive compound.

2.4.3. Mercury nitrides

Three compounds of mercury with nitrogen are known: nitride Hg_3N_2 and azides $Hg_2(N_3)_2$ and $Hg(N_3)_2$. Mercury nitride is one of the strongest explosive compounds and it is obtained by nitriding mercury oxide with ammonia at ca. 100°C:

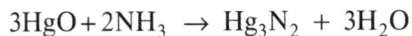

$$3HgO + 2NH_3 \rightarrow Hg_3N_2 + 3H_2O$$

or by decomposition of NHgl. Mercury azides are obtained by precipitation from solutions of corresponding nitrates by the solution of sodium azide. Both $Hg_2(N_3)_2$ and $Hg(N_3)_2$ have explosive properties.

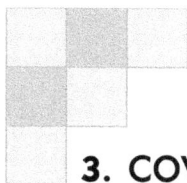

3. COVALENT NITRIDES

3.1. Nitride and oxynitride compounds of elements of the second period

3.1.1. Nitride and oxynitride compounds of boron

Boron, besides boron nitride BN, forms compounds with nitrogen of a different oxidation state of the coordination center, and also of a larger coordination number of nitrogen round boron than in BN. Fig. 4. presents the classification table of the $e_v - e_z$ axes with the placed known nitrogen compounds of boron. Fields of $e_z = 2n$ (for n = 1, 2) are only possible in the classification table for monocentric species with nitride ligands. Species lying on the $e_z = 2n$ line, except boron nitride, occur exclusively in the protonated form, i.e. in compounds with hydrogen. There are consecutive products of the condensation, to which the adduct formed from the addition of ammonia to boron hydride undergoes [8].

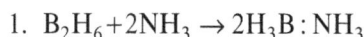

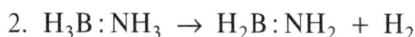

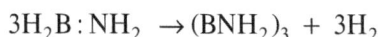

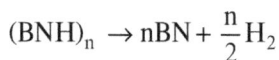

1. $B_2H_6 + 2NH_3 \rightarrow 2H_3B : NH_3$

2. $H_3B : NH_3 \rightarrow H_2B : NH_2 + H_2$

   $3H_2B : NH_2 \rightarrow (BNH_2)_3 + 3H_2$

   $n(BNH_2)_3 \rightarrow 3(BNH)_n + \dfrac{3}{2}nH_2$

   $(BNH)_n \rightarrow nBN + \dfrac{n}{2}H_2$

The numbers of the reactions correspond to those of the transformations in the classification table in Fig. 4.

$e_z\left(N^{3-}\right)$

8

7

6   BN_3^{6-}

5   4 5

4   BN_2^{3-} BN_2^{5-}

3   3

2   BN^0 $\left(BN^-\right)_n$ $\left(BN^{2-}\right)_3$ 2 $\left(BN^{4-}\right)_3$ BN^{6-}

1   1

0   B_n^0 B_2^{6-}

   0   1   2   3   4   5   6   7   8 e_v

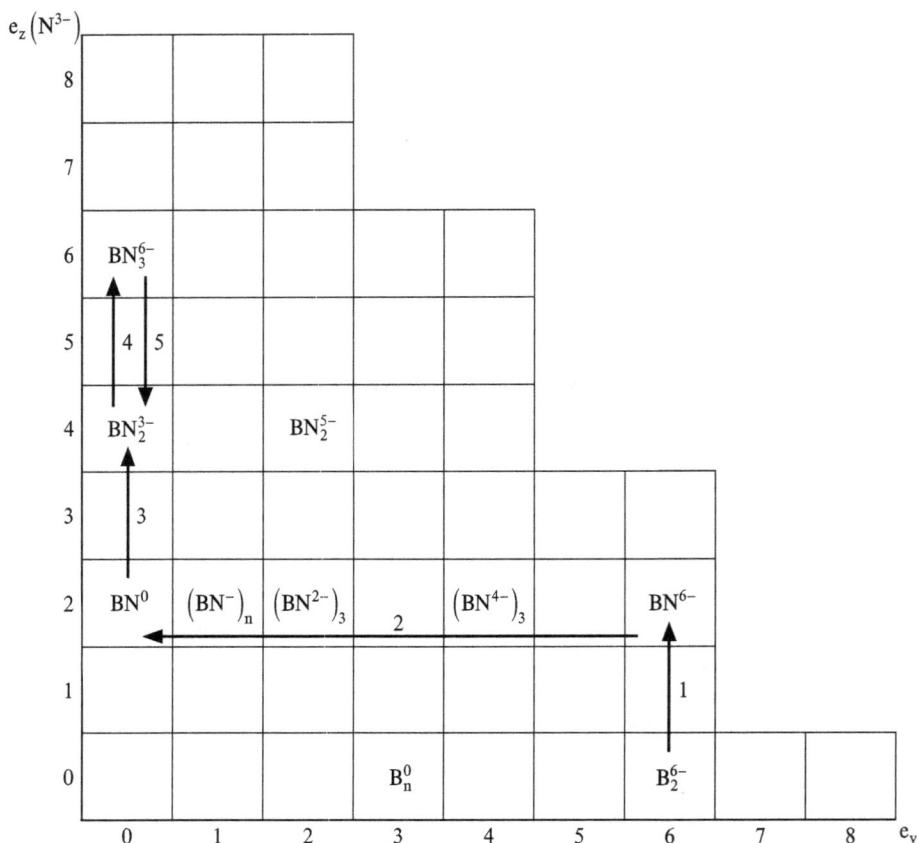

Fig. 4. Nitride compounds of boron

Species of a larger coordination number of nitrogen round boron than in boron nitride: BN_2^{3-} and BN_3^{6-} are known exclusively in the form of complex nitrides. These are salts formed in the reactions of boron nitride with lithium or beryllium, magnesium and alkaline earth metal nitrides [9].

Properties of boron nitride

Boron nitride is a compound composed of two elements of moderate electronegativity equal to that of carbon (according to Pauli), and hence results in a considerable similarity of physical and chemical properties of these two substances.

Boron nitride occurs in three crystalline forms: regular (resembling diamond), hexagonal (resembling graphite) and rhombohedral. α-BN (hexagonal form) has the lattice constants a = 2.504Å and c = 6.66lÅ; it is characterized by a very low hardness. β-BN (regular form) of the lattice constant a = 3.6lÅ shows a very high hardness (comparable with diamond). γ-BN (rhombohedral form earlier considered as hexagonal, fully packed) has the lattice constant equal to a = 2.504Å and c = 10.0lÅ.

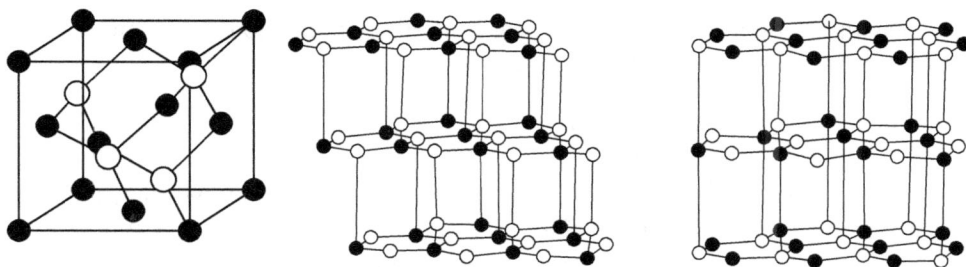

Fig. 5. The structures of boron nitrides

The hexagonal form, α-BN, can occur in a highly disordered structure, and only annealing at temperatures up to 1800°C leads to spatial developing of the lattice of the graphite type. The crystallization is of a constant character, which gives the possibility of obtaining materials of various granulation and developing crystalline lattice. Moreover, the corrosion resistance of boron nitride strongly depends on the development of the surface, ordering of the crystalline structure and level of contamination. α-BN is resistant to the treatment of reducing agents. Hydrogen up to 2000°C, fused metals up to 1500°C and many fused salts do not react with α-BN. This information is of practical importance, due to the application of boron nitride for the preparation of special ceramics, including metallurgical. However, boron nitride undergoes fast erosion above 3000°C, especially in a gas flux of high linear velocity. Boron nitride is less resistant to the action of oxidizing agents. The powder of crystalline, hexagonal BN undergoes oxidation with oxygen at ca.700°C, according to the reaction:

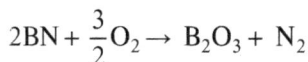

$$2BN + \frac{3}{2}O_2 \rightarrow B_2O_3 + N_2$$

and also with carbon dioxide:

$$2BN + 3CO_2 \rightarrow B_2O_3 + 3CO + N_2$$

Free fluorine and hydrogen fluoride react with BN at room temperature:

$$2BN + 3F_2 \rightarrow 2BF_3 + N_2$$

$$BN + 4HF \rightarrow NH_4BF_4$$

Chlorine causes the decomposition of BN at above 500°C:

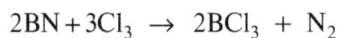

$$2BN + 3Cl_3 \rightarrow 2BCl_3 + N_2$$

The fusion of boron nitride with hydroxides leads to ammonia formation, which is the basis of one of the methods of nitrogen determination:

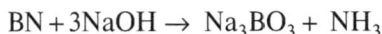

$$BN + 3NaOH \rightarrow Na_3BO_3 + NH_3$$

The reaction course of boron nitride with strong donors of nitride anions: lithium and beryllium, magnesium and alkaline earth metal nitrides (i.e. ionic nitrides) is of interest. Complex nitrides are formed in these processes, consisting of BN_2^{3-} and BN_3^{6-} in their structures [10].

3. $BN + Li_3N \rightarrow Li_3BN_2$

4. $Li_3BN_2 + Li_3N \rightarrow Li_6BN_3$

The numbers of the reactions correspond to those of the transformations in the classification table in Fig. 4.

Li_3BN_2 is of practical importance in the preparation of the regular form of boron nitride (borazon). The transformation of α-BN into the β form requires, as is known, a temperature of ca. 1500°C and pressure up to 70 kbar. However, borazon crystallizes from fused Li_3BN_2 already at a pressure of 30 kbar, i.e. under much more moderate conditions.

5. $Li_3BN_2 \rightarrow BN + Li_3N$

Salts (complex nitrides) with the BN_2^{3-} anions were obtained also with beryllium, magnesium, calcium, strontium and barium cations.

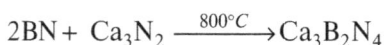

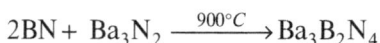

$$2BN + Ca_3N_2 \xrightarrow{800°C} Ca_3B_2N_4$$

$$2BN + Ba_3N_2 \xrightarrow{900°C} Ba_3B_2N_4$$

Preparation methods of boron nitride

On the basis of boron nitride it is possible to present all the other procedures of synthesis, in which all the other nitrides of covalent bonds of main family elements (except halogens) are obtained. Thus these methods will be precisely presented for boron in order not to repeat this description in successive chapters. The reactions described below most often lead to the formation of the α-BN form.

1. Direct synthesis from elements at about 3000°C:

$$2B + N_2 \rightarrow 2BN$$

2. Nitriding of boron compounds with ammonia

a. Nitriding of boron trioxide

The process is carried out in an ammonia flux by diluting B_2O_3 in fused alkali metal chlorides or by using solid additives such as calcium oxide, carbonate or phosphate, called carriers. Concisely, the reaction equation can be presented as follows:

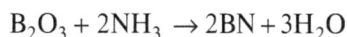

$$B_2O_3 + 2NH_3 \rightarrow 2BN + 3H_2O$$

This notion, however, does not consider many intermediate processes which occur in the case of using carriers. These are the following reactions proceeding simultaneously [11]:

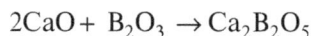

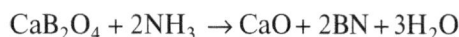

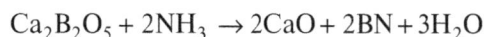

$$CaO + B_2O_3 \rightarrow CaB_2O_4$$

$$2CaO + B_2O_3 \rightarrow Ca_2B_2O_5$$

$$3CaO + B_2O_3 \rightarrow Ca_3B_2O_6$$

$$CaB_2O_4 + 2NH_3 \rightarrow CaO + 2BN + 3H_2O$$

$$Ca_2B_2O_5 + 2NH_3 \rightarrow 2CaO + 2BN + 3H_2O$$

Calcium orthoborate $Ca_3B_2O_6$ does not undergo nitriding up to 1500°C.

Ammonium chloride added directly to boron trioxide may also be a source of ammonia. Boron nitride is obtained in these processes at 800–1800°C in a mixture with borates and calcium oxide. These compounds are removed by washing out with solutions of acids. This is connected with considerable losses of boron nitride.

b. Reaction of boron trichloride or boron tribromide with ammonia in the gaseous phase, at 600–1800°C:

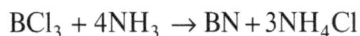

$$BCl_3 + 4NH_3 \rightarrow BN + 3NH_4Cl$$

c. Reaction of boron hydride B_2H_6 with ammonia or thermal decomposition of the $H_3B:NH_3$ adduct (the reactions are presented in the first part of this section).

In other volatile boron compounds, e.g. orthoborate acid esters may be the source of boron, which in reaction with ammonia in the gaseous phase yield boron nitride.

Solid boron phosphide in reaction with ammonia undergoes transformation into boron nitride:

$$BP + NH_3 \rightarrow BN + PH_3$$

3. Reduction of boron trioxide with simultaneous nitriding at 1500–2000°C:

$$B_2O_3 + 3C + N_2 \rightarrow 2BN + 3CO$$

This reaction occurs in low yields, and the product is contaminated with carbon or additionally with boron carbide.

4. Application of inorganic compounds containing nitrogen as nitriding agents.

These compounds can be: amides, cyanides, cyanates and cyanamides. The reactions proceeding between calcium cyanamide and boron trioxide are presented as an example:

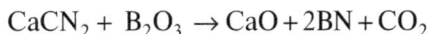

$$CaCN_2 + B_2O_3 \rightarrow CaO + 2BN + CO_2$$

The following reactions proceed in successive stages of the synthesis (also side reactions) [11]:

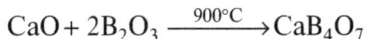

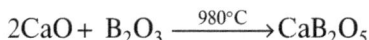

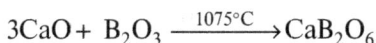

$$CaO + 2B_2O_3 \xrightarrow{900°C} CaB_4O_7$$

$$2CaO + B_2O_3 \xrightarrow{980°C} CaB_2O_5$$

$$3CaO + B_2O_3 \xrightarrow{1075°C} CaB_2O_6$$

From among the borates formed, CaB_2O_6 does not undergo nitriding, and the other two react with the cyanamide as follows:

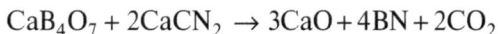

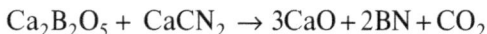

$$CaB_4O_7 + 2CaCN_2 \rightarrow 3CaO + 4BN + 2CO_2$$

$$Ca_2B_2O_5 + CaCN_2 \rightarrow 3CaO + 2BN + CO_2$$

All the calcium compounds are washed out from the post-reaction mixture with acid solutions. Crystalline, hexagonal boron nitride is obtained in up to 80% yield. The synthesis of boron nitride with the use of cyanides and amides as nitriding agents is a more complicated course. The following reactions reported in the literature on the subject:

$$B_2O_3 + 2NaCN \rightarrow 2BN + Na_2O + 2CO$$

$$B_2O_3 + 3NaNH_2 \rightarrow 2BN + NH_3 + 3NaOH$$

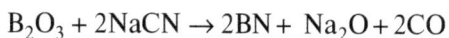

are incorrect, since, as is described in the next part of this section, boron nitride reacts violently already at 300°C both with sodium hydroxide and oxide.

5. Reactions of organic compounds containing nitrogen with boric acid or boron trioxide

Urea and products of its thermal condensation: biuret, triuret, dicyanodiamide, cyanuric acid, ammelide, ammeline, melamine and also thiourea, thiosemicarbazide and others can be used as nitriding agents. Orthoboric acid is usually mixed with an excess of the nitriding agent and is heated at an appropriate temperature (1300–1500°C) in a nitrogen or ammonia atmosphere. In some cases the carriers mentioned earlier (calcium oxide or carbonate) are used. The reaction of urea with orthoboric acid, proceeding in a few stages, is the most often applied method of BN synthesis [11]:

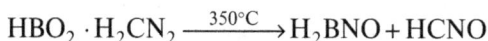

$$H_3BO_3 \rightarrow HBO_2 + H_2O$$

$$HBO_2 + H_2NCONH_2 \xrightarrow{185°C} HBO_2 \cdot H_2CN_2 + H_2O$$

$$HBO_2 \cdot H_2CN_2 \xrightarrow{350°C} H_2BNO + HCNO$$

$$H_2BNO \xrightarrow{1000°C} BN + H_2O$$

The process is carried out in an ammonia atmosphere, which permits obtaining boron nitride at 1000°C without contamination in the form of boron oxide. Melamine is also an easily available nitriding agent. Two procedures of the nitride synthesis are used with the application of melamine. The first one consist in heating a melamine mixture with orthoboric acid in a flux of ammonia up to 1200–1800°C.

The second method consist in the preparation of melamine diborate precipitating after mixing aqueous solutions of orthoboric acid and melamine. The crystallizing compound of the composition $C_3N_6H_6 \cdot 2H_3BO_3$, upon drying, is subjected to thermal decomposition in an inert atmosphere at 1200–1800°C yielding boron nitride.

6. Other, specific methods of BN synthesis

The reaction of boron hydride with hydrazine in the gaseous phase proceeding in the form of a flame:

$$B_2H_6 + N_2H_4 \rightarrow 2BN + 5H_2$$

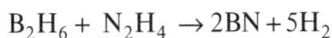

Reaction of magnesium nitride with free boron at 1300°C:

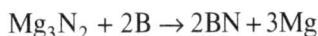

$$Mg_3N_2 + 2B \rightarrow 2BN + 3Mg$$

The latter process proceeding under appropriately high pressure yields the regular form of boron nitride (borazone).

Oxynitride compounds of boron

An amorphous phase of the H_2BNO composition is one of the transition products formed during the synthesis of boron nitride from the reaction of orthoboric acid with urea (this is precisely described in the first section of this chapter). This intermediate product indicates the possibility of the existence of the BNO^{2-} anion, species of a mixed oxynitride surrounding round the coordination center (boron). It seems sensible to present here the application of the morphological classification of simple species to predict the existence of new chemical compounds. Among the carbon compounds salts of a mixed oxynitride surrounding of carbon: cyanates (CNO^- anions) and carbaminates (CNO_2^{3-}), have been known for a long time. Thus, it could not be excluded that boron, adjacent to carbon in the same period, can form isoelectronic species BNO^{2-} or BNO_2^{4-}. The answer to the question whether it is possible to obtain salts containing BNO^{2-} anions, was obtained during studies on the course of the reaction of boron nitride with lithium and sodium oxides [11]. The hexagonal form of boron nitride undergoes an exothermic reaction with lithium oxide at 270°C. A new, formerly unknown compound of the Li_2BNO composition is the product of this process. Also sodium oxide in a strongly exothermic reaction with boron nitride at 270°C forms a new, not described previously, salt Na_2BNO.

In Fig. 6. of the $e_z(O^{2-}) - e_z(N^{3-})$ coordinate system the oxide, nitride and hypothetic oxynitride species of boron are presented. The directions of chemical transformations are also marked: synthesis and thermal decomposition of new salts. As it is known, cyanides undergo thermal decomposition with the formation of two products: a compound of a purely nitride coordination surrounding and a compound of a purely oxide surrounding the coordination center:

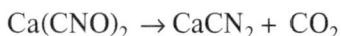

$$Ca(CNO)_2 \rightarrow CaCN_2 + CO_2$$

If the analogies between the boron and carbon compounds also concerned the course of the thermal decomposition, then the oxynitride-borates should transform into the borate and a compound of a purely nitride surrounding the coordination center. Experimental studies showed that the course of both processes is as follows (the numbers of the reactions correspond to those of the transformations in Fig. 6.):

42

1. $BN + Li_2O \rightarrow Li_2BNO$

   $BN + Na_2O \rightarrow Na_2BNO$

   $BN + O^{2-} \rightarrow BNO^{2-}$

2. $2Li_2BNO \rightarrow Li_3BN_2 + LiBO_2$

   $2BNO^{2-} \rightarrow BN_2^{2-} + BO_2^-$

3. $3Na_2BNO \rightarrow 2BN + Na_3BO_3 + Na_3N$

   $3BNO^{2-} \rightarrow 2BN + BO_3^{3-} + N^{3-}$

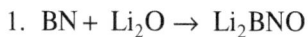

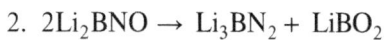

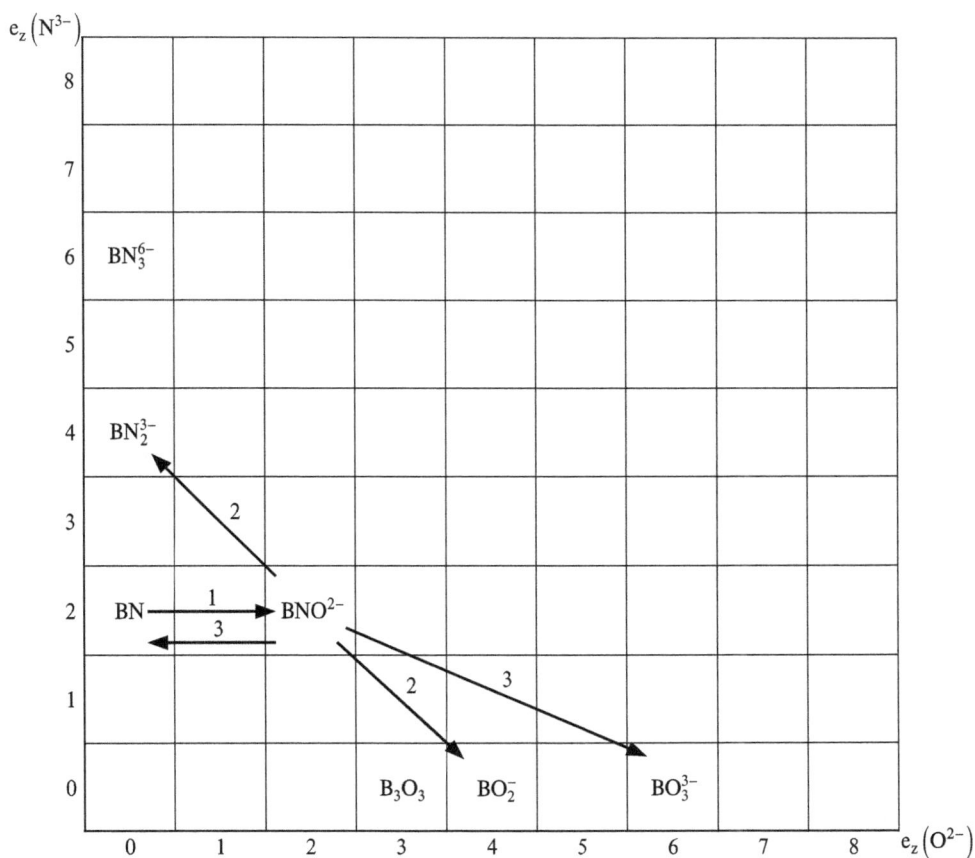

Fig. 6. Oxynitride compounds of boron

The application of the morphological classification of simple species for the prediction of ways of preparation and direction of the thermal decomposition of hypothetic fluoroazo compounds will be shown. In Fig. 7. the known fluoride and nitride species of boron, as well as species of a mixed fluoronitride coordination surrounding, not obtained till now, are presented in the $e_z(F^-) - e_z(O^{2-})$ coordinate system. The preparation methods of one of them are proposed:

1. $BN + 3F^- \rightarrow 2BNF_3^{3-}$

2. $BF_3 + N^{3-} \rightarrow 2BNF_3^{3-}$

This problem is discussed in more details in a separate chapter.

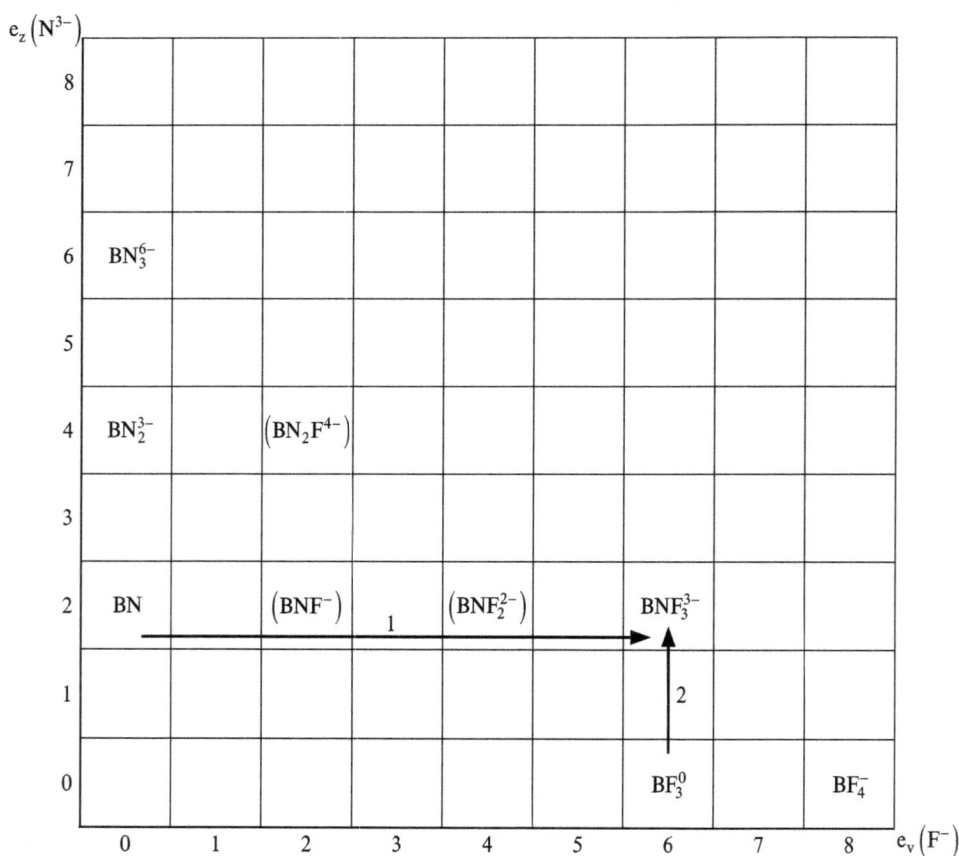

Fig. 7. Fluoronitride compounds of boron

3.1.2. Nitride and oxynitride compounds of carbon

Carbon nitrides

Carbon at the maximum oxidation state forms the nitride C_3N_4. It has a hexagonal structure and hardness like that of diamond [12]. It is obtained from the elements in the synthesis at low temperature plasma under reduced pressure of nitrogen:

$$3C + 2N_2 \rightarrow C_3N_4$$

C_2N_2, dicyan, is also known.

Similarly as boron, carbon forms species – anions, in which the coordination center is surrounded by a greater number of nitride ligands: CN_2^{2-} and CN_3^{5-}.

CN_2^{2-} occurs both in the form of an anion in salts – cyanamides, and in the protonated form in H_2CN_2 – cyanamide. On the other hand, CN_3^{5-} is known only in the protonated form in guanidine – H_5CN_3. The cyanide anions occurs both in the form of a salt and in the protonated as hydrogen cyanide.

Properties of carbon nitrides

The structures of species placed in the table in Fig. 8. are presented below. From the example of CN_2^{2-} the influence of the type of cationic counterion on the length of the carbon-nitrogen bond is clearly visible (in Å).

hydrogen cyanide HCN $H \xrightarrow{1,06Å} C \xrightarrow{1,15Å} N$

cyanamide H_2CN_2 $H_2N \xrightarrow{1,31Å} C \xrightarrow{1,15Å} N$

lead cyanamide $PbCN_2$ $N \xrightarrow{1,25Å} C \xrightarrow{1,17Å} N^{2-}$

calcium cyanamide $N \xrightarrow{1,22Å} C \xrightarrow{1,22Å} N^{2-}$

dicyanide $N \xrightarrow{1,15Å} C \xrightarrow{1,38Å} C \xrightarrow{1,15Å} N$

Carbon nitrides were omitted in the monographs devoted to the nitrides probably due to the fact that their properties differ from the typical compound properties of the other elements with nitrogen. Hydrogen cyanide and dicyanide are gases, guanidine is an unstable compound, stable in the form of the $H_6CN_3^+$ cation occurring in salts. Cyanamides with the CN_2^{2-} anion also have not been treated as complex nitrides. The chemical transformations linking the particular carbon nitrides placed in the table in

Fig. 8. will be shortly presented. The numbers of the reactions correspond to those of the transformations in the figure.

1. $HCN + Cl_2 \rightarrow CNCl + HCl$

2. $CNCl + 2NH_3 \rightarrow H_2CN_2 + NH_4Cl$

3. $Ca(CN)_2 \xrightarrow{1000°C} CaCN_2 + C$

4. $Hg(CN)_2 \rightarrow Hg + (CN)_2$

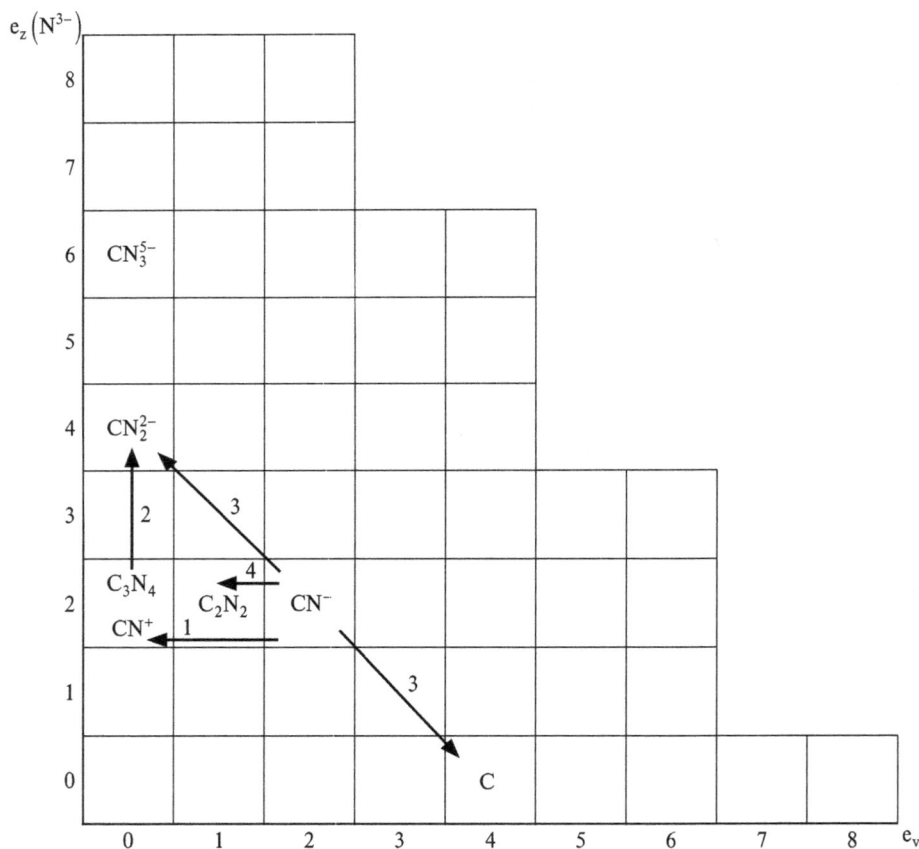

Fig. 8. Nitride compounds of carbon

Carbon oxynitrides

Carbon forms a considerable number of compounds in which it has a mixed oxynitride coordination surrounding. By taking advantage of the classification table in the

$e_z(O^{2-}) - e_z(N^{3-})$ coordinate system it is possible to present the transformations linking monocentric species. The presentation of all of the oxynitride carbon species would require a three-dimensional classification space $e_z(O^{2-}) - e_z(N^{3-}) - n$, where n denotes the number of carbon atoms in the species, i.e. the number of coordination centers. By introducing a new parameter, number $\bar{e}_z$ – number of electrons formally introduced by all the ligands in the species corresponding to one coordination center, it is possible to simplify the system to two dimensions and to present it on a plane. Carbon oxynitrides are formed during the thermal condensation of urea in a temperature of up to 300°C, the course of which, with some simplifications, is presented by the following reactions [11]:

1. $CO(NH_2)_2 \rightarrow HCNO + NH_3$

2. $CO(NH_2)_2 \rightarrow H_2CN_2 + H_2O$

3. $2H_2CN_2 \rightarrow H_4C_2N_4$

4. $3H_2CN_2 \rightarrow H_6C_3N_6$

5. $CO(NH_2)_2 + HCNO \rightarrow H_2NCONHCONH_2$

6. $3HCNO \rightarrow (HCNO)_3$

The following compounds are formed:

HCNO – cyanic acid ($\bar{e}_z = 4$)

H_2CN_2 – cyanamide ($\bar{e}_z = 4$)

$H_4C_2N_4$ – dicyandiamide ($\bar{e}_z = 4$)

$H_6C_3N_6$ – melamine ($\bar{e}_z = 4$)

$H_2NCONHCONH_2$ – biuret ($\bar{e}_z = 5$)

$(HCNO)_3$ – cyanuric acid ($\bar{e}_z = 4$)

Fig. 9. presents the classification table in the $n - \bar{e}_z$ system with the protonless skeletons of urea and the products of its thermal condensation and directions of transformations.

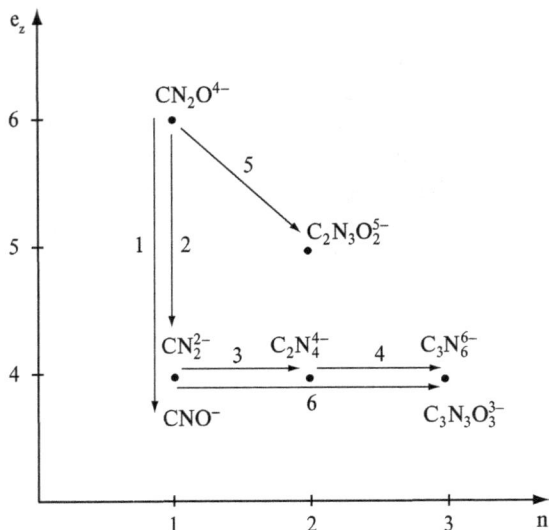

Fig. 9. Oxynitride compounds of carbon

3.2. Nitride and oxynitride compounds of elements of the third period

3.2.1. Nitride and oxynitride compounds of aluminum

Aluminum, like aluminum nitride AlN, also forms salts with the AlN_2^{3-} anions. Fig. 10 presents the classification table in the $e_v - e_z$ coordinate system with the placed nitride species. Aluminum azide $Al(N_3)_3$ is also known.

3.2.1.1. Properties of aluminum nitride

Aluminum nitride occurs in a hexagonal, completely packed crystalline structure of the wurtzite type. It is characterized by high thermal stability; its melting point (with decomposition) in non-oxidative media is 2200–2500°C. Crucibles for fusing metals, ceramic packings operating at up to 1500°C in highly aggressive reducing media and plates for the setting of semiconductive elements are made from aluminum nitride. AlN is quite stable chemically, but the fine powders obtained directly from the synthesis easily hydrolyze in water. The material becomes resistant to strongly corrosive agents of an acidic and reducing character only upon heating (e.g. 1800°C) or transformation during sintering into profiles. Aluminum nitride powder annealed at a high temperature, i.e. stabilized, does not react with oxygen, nor (up to 750°C) with chlorine. The following reactions take place above those temperatures:

48

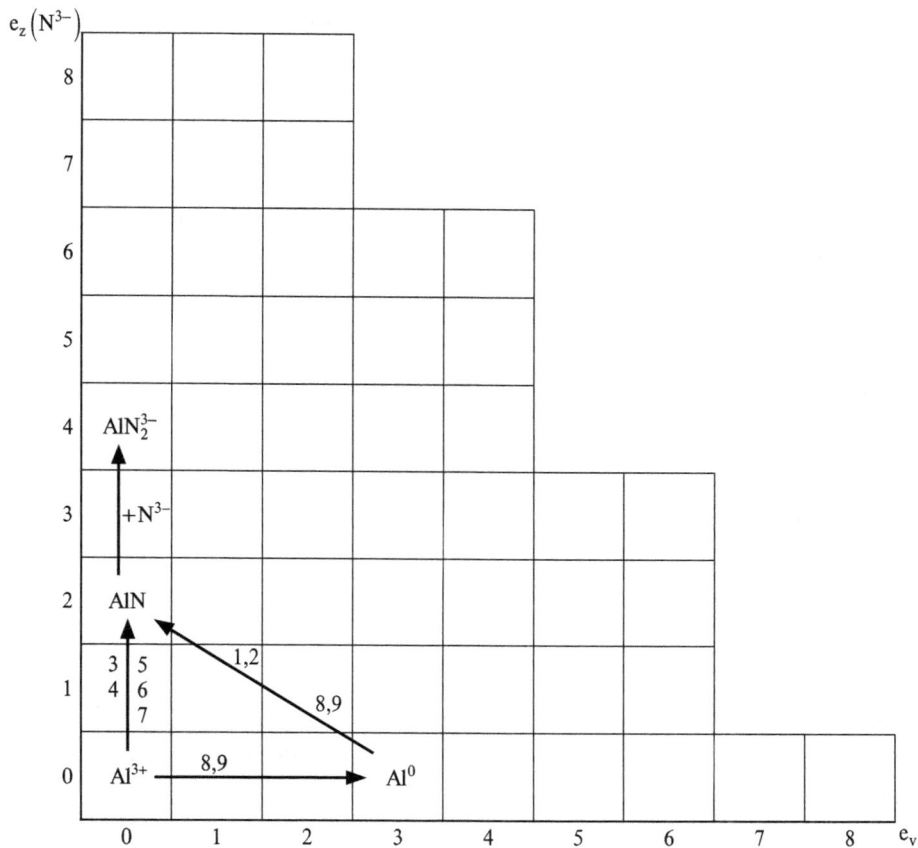

Fig. 10. Nitride compounds of aluminum

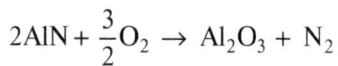

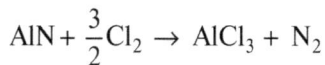

$$2AlN + \frac{3}{2}O_2 \rightarrow Al_2O_3 + N_2$$

$$AlN + \frac{3}{2}Cl_2 \rightarrow AlCl_3 + N_2$$

Concentrated, hot aqueous solutions of sodium or potassium hydroxides cause slow decomposition of aluminum nitride:

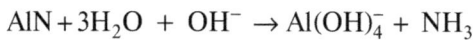

$$AlN + 3H_2O + OH^- \rightarrow Al(OH)_4^- + NH_3$$

In order to transform aluminum nitride into compounds soluble in water to carry out quantitative analysis of its content, it is fused with solid lithium or sodium hydroxides:

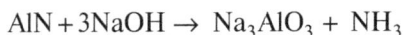

$$AlN + 3NaOH \rightarrow Na_3AlO_3 + NH_3$$

Above 1200°C aluminum nitride reacts with carbon with the formation of carbide. Contrary to boron nitride, it does not reduce carbon oxides up to 1800°C – probably due to the formation of a passive layer of aluminum oxycarbide. Fused gallium arsenide or indium antimonide do not react with aluminum nitride, hence crucibles for the preparation of single crystals of these compounds are produced from AlN. Aluminum nitride in reactions with strong donors of nitride anions – ionic lithium, magnesium, calcium, strontium or barium nitrides forms complex nitrides, in which the anionic sublattice is formed by AlN_2^{3-} species:

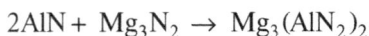

$$AlN + Li_3N \rightarrow Li_3AlN_2$$

$$2AlN + Mg_3N_2 \rightarrow Mg_3(AlN_2)_2$$

$$AlN + N^{3-} \rightarrow AlN_2^{3-}$$

This transformation is marked in the classification table in Fig. 10.

Lithium-aluminum nitride can be obtained by direct nitriding a mixture of lithium and aluminum at 700°C:

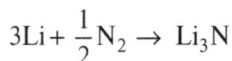

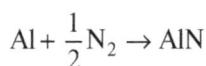

$$3Li + \frac{1}{2}N_2 \rightarrow Li_3N$$

$$Al + \frac{1}{2}N_2 \rightarrow AlN$$

$$Li_3N + AlN \rightarrow Li_3AlN_2$$

or by nitriding Li_3Al with nitrogen:

$$Li_3Al + N_2 \rightarrow Li_3AlN_2$$

3.2.1.2. Methods of obtaining of aluminum nitride

In comparison with the methods of boron nitride synthesis presented in the previous section, in the case of aluminum, at elevated temperatures in which nitriding occurs, aluminum occurs as the liquid phase (boron trioxide for boron) and aluminum trioxide as the solid phase (free boron in the other case). Hence the different results among the methods of obtaining both nitrides.

1. Direct synthesis from elements

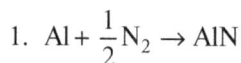

1. $Al + \frac{1}{2}N_2 \rightarrow AlN$

It is possible to perform the process in different manners. Fine aluminum powders are nitrided at temperatures below the melting point of the metal. The contact surface of the gas phase with the nitrided metal decreases upon fusing aluminum. The pulverization of the fused aluminum into vapor in a nitrogen flux at 800–2000°C is another solution. In all these methods it is expedient to carry out the reaction under high pressure of nitrogen.

2. Ammonia as the nitriding agent

Ammonia undergoes a reaction with metallic aluminum yielding a nitride with hydrogen evolution:

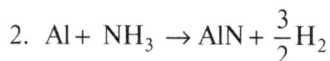

2. $Al + NH_3 \rightarrow AlN + \frac{3}{2}H_2$

A compact layer of aluminum nitride is formed on the surface of the metal (both fused and in the solid state), considerably limiting the diffusion of nitrogen or ammonia, similarly as when nitriding aluminum with nitrogen. Only at above 950°C the rate of diffusion, and thus of nitriding, is sufficient for applying this method industrially. Aluminum trioxide is transformed into a nitride in reaction with ammonia:

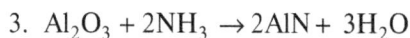

3. $Al_2O_3 + 2NH_3 \rightarrow 2AlN + 3H_2O$

The reaction of aluminum trichloride with ammonia in the gaseous phase at 800–1500°C leads to aluminum nitride. This method permits depositing layers of the nitride on bases placed in the reaction zone:

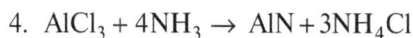

4. $AlCl_3 + 4NH_3 \rightarrow AlN + 3NH_4Cl$

This process can be carried out in another way. At room temperature ammonia forms an adduct with aluminum chloride, the decomposition of which at 800°C to 1800°C leads to the formation of aluminum nitride with hydrogen chloride evolution:

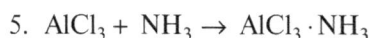

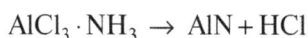

5. $AlCl_3 + NH_3 \rightarrow AlCl_3 \cdot NH_3$

$AlCl_3 \cdot NH_3 \rightarrow AlN + HCl$

Other volatile aluminum compounds can also be applied in the reaction with ammonia (at 300–800°C):

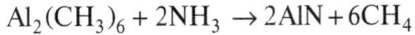

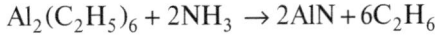

$$Al_2(CH_3)_6 + 2NH_3 \rightarrow 2AlN + 6CH_4$$

$$Al_2(C_2H_5)_6 + 2NH_3 \rightarrow 2AlN + 6C_2H_6$$

The thermal decomposition of ammonium-aluminum fluoride leads to aluminum nitride (in low yield):

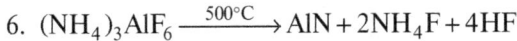

6. $(NH_4)_3AlF_6 \xrightarrow{\;500°C\;} AlN + 2NH_4F + 4HF$

In this reaction also ammonia is the nitriding agent, however, in the form of an ammonium cation. Aluminum phosphide transforms into nitride under the action of ammonia:

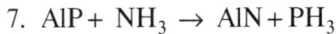

7. $AlP + NH_3 \rightarrow AlN + PH_3$

3. Reduction of aluminum trioxide with simultaneous nitriding:

8. $Al_2O_3 + 3C + N_2 \rightarrow 2AlN + 3CO$

Aluminum nitride obtained by this method is considerably contaminated with carbon and aluminum carbide as well as aluminum azocarbides. A smaller amount of this type of impurities is found in aluminum nitride, when methane is used as the reducing agent:

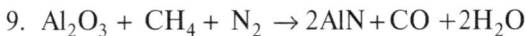

9. $Al_2O_3 + CH_4 + N_2 \rightarrow 2AlN + CO + 2H_2O$

The numbers of the successive syntheses of aluminum nitride correspond to those of the respective transformations presented in the classification table in Fig. 10.

3.2.1.3. Oxynitride compounds of aluminum

γ-AlON phase

The crystalline phase known as γ-AlON is of a spinel structure. The composition of this phase does not correspond to the simple Al_3O_3N stoichimetry, but is similar to that expressed by the formula $9Al_2O_3 \cdot 5AlN$ [13]. The interest shown in aluminum oxynitride results from the exceptional properties of the AlON phases. They fuse above 2050°C, and the glass obtained of the composition $9Al_2O_3 \cdot 5AlN$ transmits radiation within the whole infrared, visible and ultraviolet ranges.

Oxynitride salts of aluminum

The exothermal reaction of aluminum nitride with lithium oxide results in the formation of a crystalline compound of the Li_2AlNO composition:

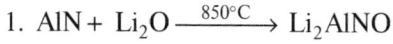

1. $AlN + Li_2O \xrightarrow{850°C} Li_2AlNO$

The same compound is formed in the reaction of aluminum trioxide with lithium nitride but first a double-exchange reaction proceeds:

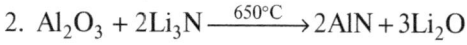

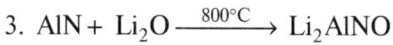

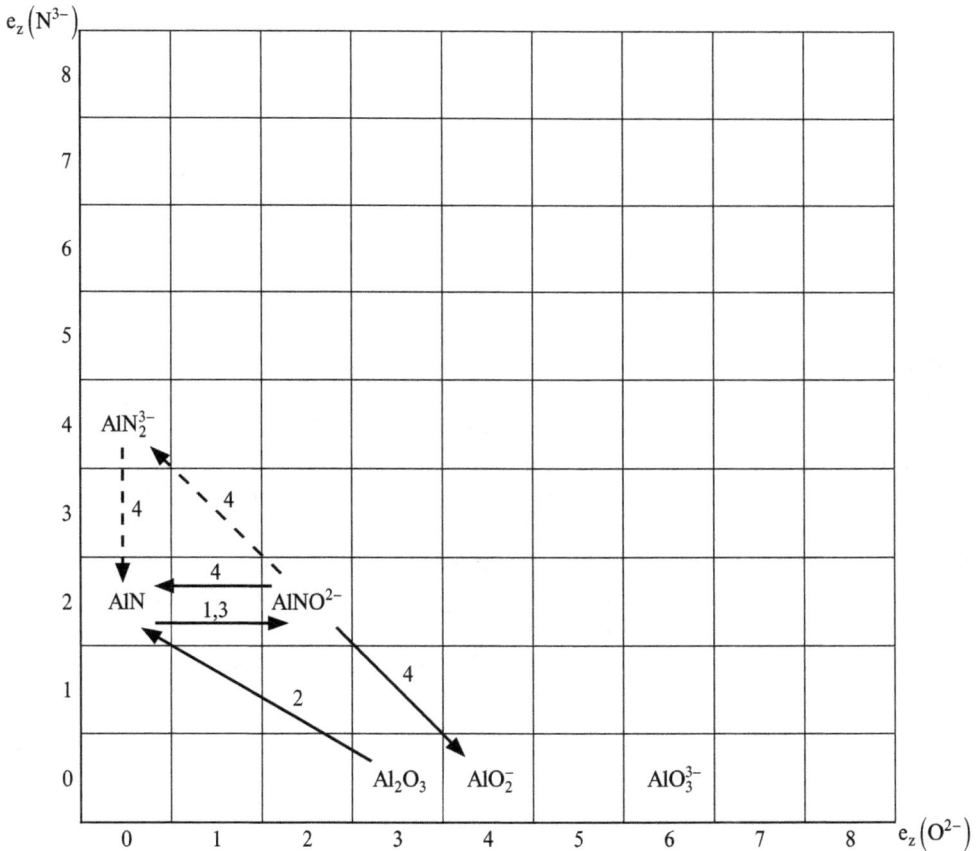

2. $Al_2O_3 + 2Li_3N \xrightarrow{650°C} 2AlN + 3Li_2O$

3. $AlN + Li_2O \xrightarrow{800°C} Li_2AlNO$

Fig. 11. Oxynitride compounds of aluminum

The compound formed of a mixed oxynitride coordination surrounding undergoes thermal decomposition in the direction of two compounds of purely oxide and nitride surrounding of the coordination center:

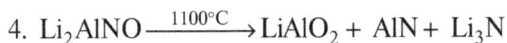

$$4. \ Li_2AlNO \xrightarrow{1100°C} LiAlO_2 + AlN + Li_3N$$

In Fig. 11. the classification table of the $e_z(O^{2-}) - e_z(N^{3-})$ axes with marked oxide, nitride and oxonitride species of aluminum is presented. The numbers of the above reactions correspond to those of the transformations in the table.

3.2.2. Nitride and oxynitride compounds of silicon

Silicon, in comparison with boron or aluminum, forms a considerably greater number of compounds with nitrogen than the ligand. Besides silicon nitride – the compound of the widest practical application among the ones described here, a number of species at the maximum oxidation degree are known, occurring both in salts and in the protonated form. These are: $Si_2N_3^-$, SiN_2^{2-}, SiN_3^{5-}, SiN_4^{8-}, lying on the $e_z = 0$ line in the classification table presented in Fig. 12. The reactions involving these species are presented in a further section of this chapter.

The SiN^+ cation occurs in the compound with fluorine and the $Si_2N_4^{6-}$, SiN_2^{4-} and SiN^{3-} are known exclusively in compounds with hydrogen, formed in the reaction of silane with ammonia.

3.2.2.1. Properties of silicon nitride

This section of the chapter will be devoted not only to silicon nitride but also to complex nitrides, in which silicon is the coordination center. The compounds of silicon with nitrogen and hydrogen (amides and imides) will be considered in the next section devoted to the methods of nitride synthesis. In those reactions these compounds are intermediates. Silicon nitride Si_3N_4 occurs in a hexagonal crystalline structure in two forms: α and β. Its very high chemical stability is characteristic. Si_3N_4 is the most resistant nitride towards corrosion in any medium, also at high temperatures. The hardness and bending strength up to 1200°C place this compound among the most valuable ceramic materials. Products from silicon nitride, especially modified with additives of the β phase, find application as cutting tools, turbine blades or engine parts. The structure of β-Si_3N_4 is shown in Fig. 13.

As was mentioned, silicon nitride is the most chemically stable nitride. Si_3N_4 powder, in the air, reacts with oxygen above 1300°C:

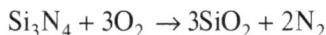

$$Si_3N_4 + 3O_2 \rightarrow 3SiO_2 + 2N_2$$

Free chlorine reacts with silicon nitride above 900°C:

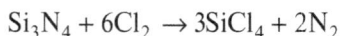

$$Si_3N_4 + 6Cl_2 \rightarrow 3SiCl_4 + 2N_2$$

54

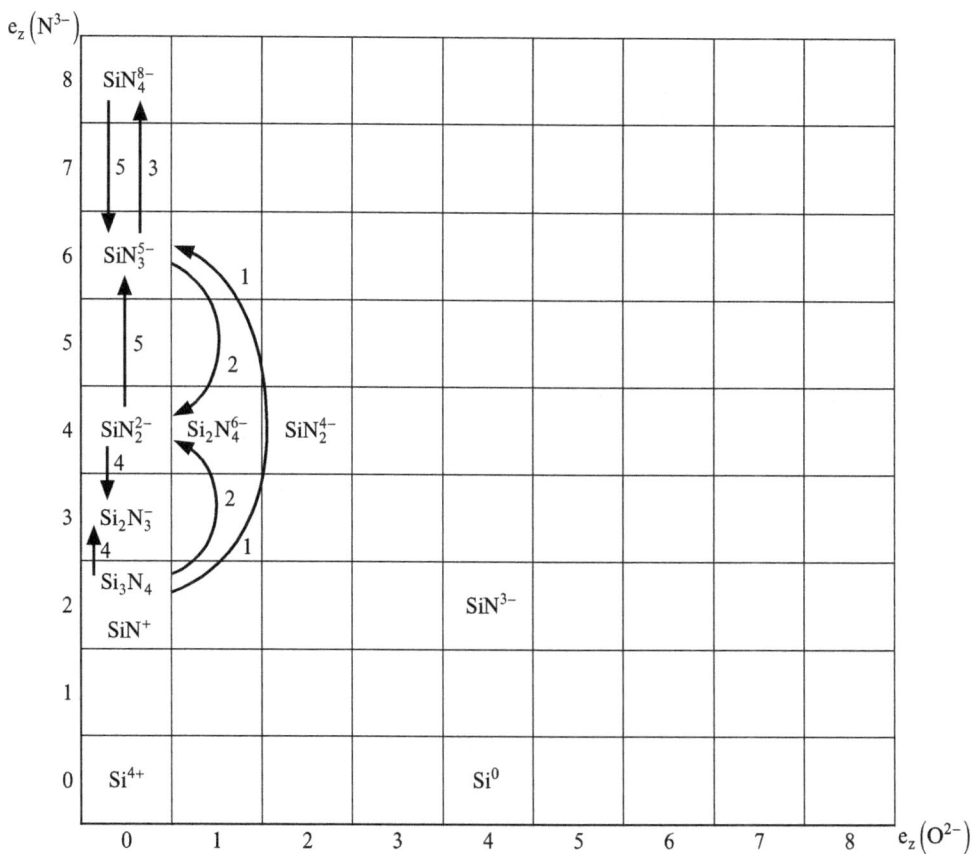

Fig. 12. Nitride compounds of silicon

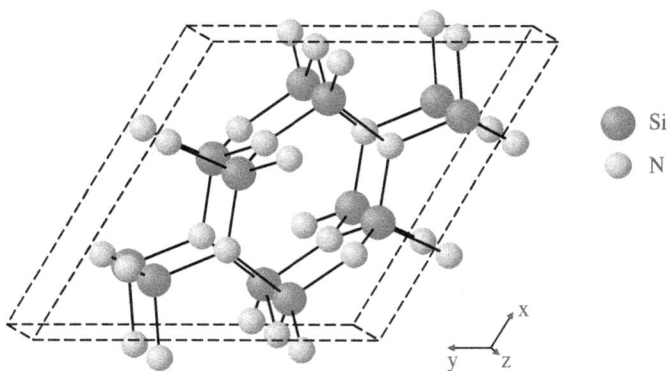

Fig. 13. The structure of β-Si_3N_4

Silicon nitride in reaction with lithium nitride forms the following compounds (the numbers of the reactions correspond to those of the transformations in Fig. 12.):

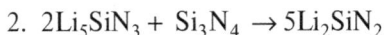

1. $Si_3N_4 + 5Li_3N \rightarrow 3Li_5SiN_3$

$Si_3N_4 + 5N^{3-} \rightarrow 3SiN_3^{5-}$

2. $2Li_5SiN_3 + Si_3N_4 \rightarrow 5Li_2SiN_2$

$2SiN_3^{5-} + Si_3N_4 \rightarrow 5SiN_2^{2-}$

3. $Li_5SiN_3 + Li_3N \rightarrow Li_8SiN_4$

$SiN_3^{5-} + N^{3-} \rightarrow SiN_4^{8-}$

4. $Li_2SiN_2 + Si_3N_4 \rightarrow 2LiSi_2N_3$

$SiN_2^{2-} + Si_3N_4 \rightarrow 2Si_2N_3^{-}$

5. $Li_2SiN_2 + Li_8SiN_4 \rightarrow 2Li_5SiN_3$

$SiN_2^{2-} + SiN_4^{8-} \rightarrow 2SiN_3^{5-}$

Complex lithium-silicon nitrides are not very chemically stable compounds. They undergo slow hydrolysis in water and in mineral acids media rapid decomposition. Complex nitrides, salts with silicon as the coordination center, are also known with beryllium, magnesium, strontium and barium cations. Beryllium nitride reacts with silicon nitride only at 1800°C:

$Be_3N_2 + Si_3N_4 \rightarrow 3BeSiN_2$

This compound is chemically stable and does not react with aqueous solutions of acids and bases. Magnesium nitride in reaction with silicon nitride also yields a complex nitride:

$Mg_3N_2 + Si_3N_4 \rightarrow 3MgSiN_2$

Both the beryllium and magnesium salts react with oxygen at high temperatures:

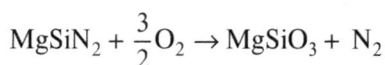

$MgSiN_2 + \frac{3}{2}O_2 \rightarrow MgSiO_3 + N_2$

Calcium nitride, similarly as lithium nitride, in reaction with silicon nitride forms salts containing: $Si_2N_3^-$, SiN_2^{2-}, SiN_3^{5-} and SiN_4^{8-}. At 600°C $Ca_5Si_2N_6$ is formed and at 800°C $CaSiN_2$ and Ca_4SiN_4.

As was mentioned earlier, species with nitride ligands, in which silicon is the coordination center, are also known in compounds with hydrogen. In the reaction of silicon tetrachloride with ammonia, which yields silicon nitride as the final product when carried out at 0°C to 1200°C, silicon amide and imide are the intermediates.

3.2.2.2. Methods of silicon nitride preparation

The methods of silicon nitride synthesis are, as far as the kind of reactants used is concerned, similar to the aluminum nitride obtaining methods.

1. Direct synthesis from elements

$$3Si + 2N_2 \rightarrow Si_3N_4$$

A large number of catalysts are used to increase the reaction rate. These are both metals (Li, Ca), nitrides (Li_3N, Ca_3N_2) as well as oxides and fluorides.

2. Nitriding silicon compounds with ammonia

The reaction of silicon tetrachloride with ammonia in the gaseous phase at above 1200°C leads directly to silicon nitride:

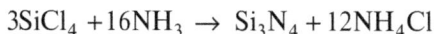

$$3SiCl_4 + 16NH_3 \rightarrow Si_3N_4 + 12NH_4Cl$$

Also another thermal procedure can be applied. First silicon tetraamide is obtained at a low temperature, which during thermal decomposition yields silicon nitride:

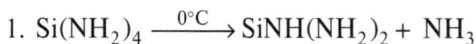

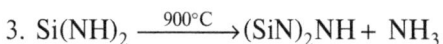

$$SiCl_4 + 8NH_3 \rightarrow Si(NH_2)_4 + 4NH_4Cl$$

1. $Si(NH_2)_4 \xrightarrow{\ 0°C\ } SiNH(NH_2)_2 + NH_3$

2. $SiNH(NH_2)_2 \xrightarrow{\ 0°C\ } Si(NH)_2 + NH_3$

3. $Si(NH)_2 \xrightarrow{\ 900°C\ } (SiN)_2NH + NH_3$

4. $3(SiN)_2NH \xrightarrow{\ 1200°C\ } 2Si_3N_4 + NH_3$

The condensation of silicon tetraamide can be presented for protonless skeletons of consecutive intermediates:

1. $SiN_4^{8-} \rightarrow SiN_3^{5-} + N^{3-}$

2. $SiN_3^{5-} \rightarrow SiN_2^{2-} + N^{3-}$

3. $2SiN_2^{2-} \rightarrow Si_2N_3^- + N^{3-}$

4. $3Si_2N_3^- \rightarrow 2Si_3N_4 + N^{3-}$

The transformations above are presented in the classification table in Fig. 14.

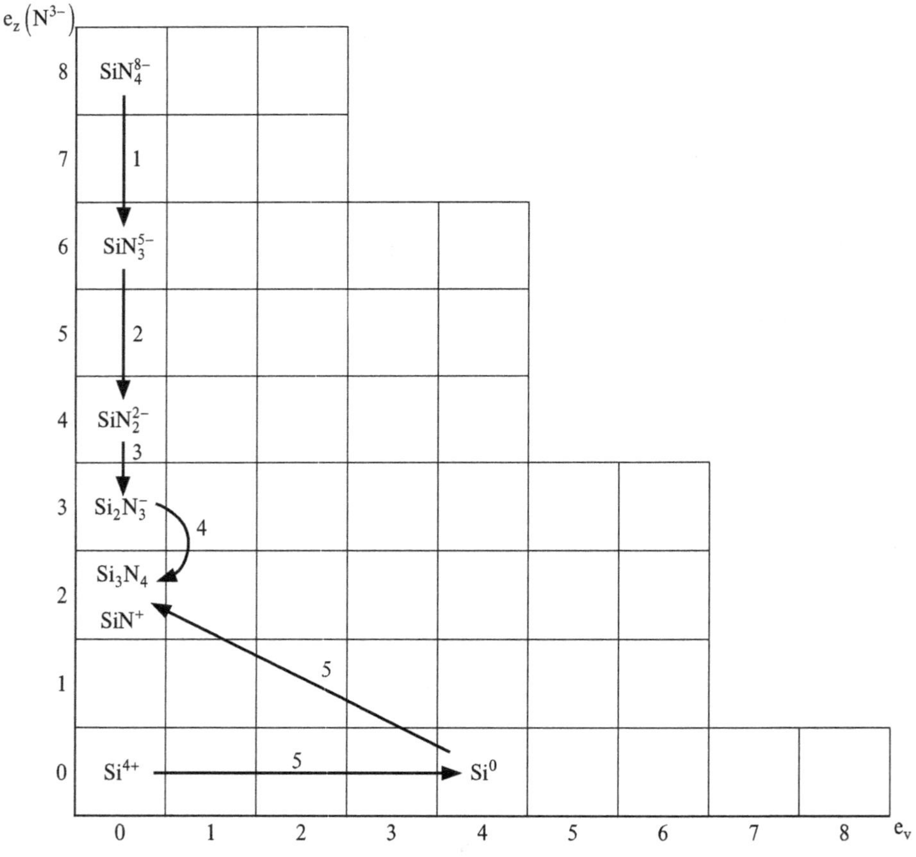

Fig. 14. The synthesis of silicon nitride compounds

It appears that similarly to the synthesis of complex lithium-silicon nitrides (Fig. 12.) consisting of simple anionization with nitride anions, the decomposition of silicon tetraamide consists of four successive deanionization stages.

Silane in reaction with ammonia in the gaseous phase yields silicon nitride and hydrogen as the final products:

$$3SiH_4 + NH_3 \rightarrow Si_3N_4 + 12H_2$$

Silicon nitride is also formed in the reactions of silane with hydrazine or tetramethylsilane with ammonia.

3. Reduction of silicon dioxide by carbon with simultaneous nitriding

$$5.\ 3SiO_2 + 3C + 2N_2 \rightarrow Si_3N_4 + 3CO_2$$

The pathway of this reaction is marked by number 5 in Fig. 14. The product is contaminated by carbon and silicon carbides.

It appeared, however, that one of the forms of the nitride, β-Si_3N_4, undergoes a reaction with oxides of some metals, bonding them without a change in its hexagonal structure and forming crystalline phases of properties similar to those of silicon nitride, but much easier to obtain by sintering. The most interesting properties are exhibited by the product of the reaction of Si_3N_4 and aluminum oxide. Due to the structural similarity to β-Si_3N_4, this phase was called β'-SiAlON. Its composition can be described by the relationship $Si_{6-z}Al_zO_zN_{8-z}$ and it is stable within $0<z<4.2$. The preparation of β'-SiAlON requires selecting an appropriate reactant mixture and then sintering it at 1500–1800°C [14, 15].

3.2.2.3. Oxynitride compounds of silicon

The number of known complex nitrides with silicon (nitride silicates) is larger than of nitrogen aluminates or nitrogen borates. Hence, a larger number of silicon compounds of a mixed coordination surrounding – oxynitride, occurs. The classification table (Fig. 15.) of the $e_z(O^{2-}) - e_z(N^{3-})$ coordinate system presents the known monocentric oxy-silicates, nitride silicates, and already obtained oxynitride silicates.

It is interesting to explain the role of the morphological classification in the process of obtaining new oxynitride silicates. On the basis of general premises the course of a number of anionization reactions was assumed, in which from known compounds ones not obtained till now were to be formed. Studies were performed in such a way so as to obtain from two different pairs of reactants the same new, unknown till now, compound. A number of processes are presented below, in which new species of a mixed oxynitride ligand surrounding round the coordination center can be formed (by the example of lithium salts).

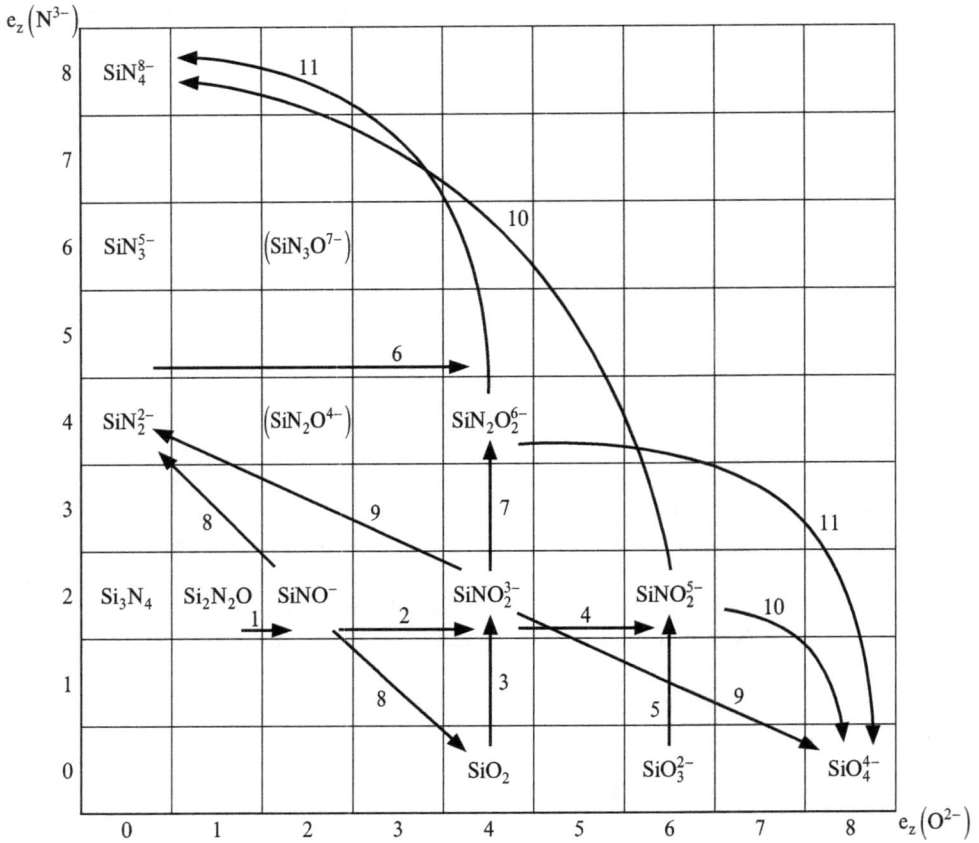

Fig. 15. Oxynitride compounds of silicon

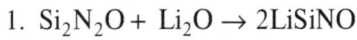

1. $Si_2N_2O + Li_2O \rightarrow 2LiSiNO$

   $Si_2N_2O + O^{2-} \rightarrow 2SiNO^-$

2. $LiSiNO + Li_2O \rightarrow Li_3SiNO_2$

   $SiNO^- + O^{2-} \rightarrow SiNO_2^{3-}$

3. $SiO_2 + Li_3N \rightarrow Li_3SiNO_2$

   $SiO_2 + N^{3-} \rightarrow SiNO_2^{3-}$

4. $Li_3SiNO_2 + Li_2O \rightarrow Li_5SiNO_3$

$$SiNO_2^{3-} + O^{2-} \rightarrow SiNO_3^{5-}$$

5. $Li_2SiO_3 + Li_3N \rightarrow Li_5SiNO_3$

$$SiO_3^{2-} + N^{3-} \rightarrow SiNO_3^{5-}$$

6. $Li_2SiN_2 + 2Li_2O \rightarrow Li_6SiN_2O_2$

$$SiN_2^{2-} + 2O^{2-} \rightarrow SiN_2O_2^{6-}$$

7. $Li_3SiNO_2 + Li_3N \rightarrow Li_6SiN_2O_2$

$$SiNO_2^{3-} + N^{3-} \rightarrow SiN_2O_2^{6-}$$

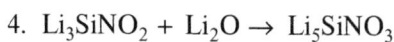

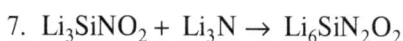

The numbers of the reactions correspond to those of the transformations in Fig. 15.

Experimental studies were carried out, in which all the plane reactions were performed. Applying complex methods of thermal analysis, X-ray phase analysis and classical chemical analysis the hypothetic oxynitride silicates in lithium and sodium salts were obtained and then identified.

However, until now the salts with SiN_2O^{4+} and SiN_3O^{7-} anions have not been obtained, but their existence seems to be probable. As has been presented in the previous chapters, the oxynitride salts of boron, aluminum and carbon transform in the thermal decomposition in the direction of two compounds: one of a purely oxide coordination surrounding round the central element and the second of a purely nitride ligand surrounding. These processes proceed according to the following reactions:

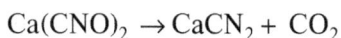

$$2Li_2BNO \rightarrow LiBO_2 + Li_3BN_2$$

$$Ca(CNO)_2 \rightarrow CaCN_2 + CO_2$$

$$Li_2AlNO \rightarrow LiAlO_2 + AlN + Li_3N$$

It was assumed that also the newly obtained oxynitride silicates would undergo decomposition in a similar route. The results of experimental studies fully confirmed this hypothesis. The thermal decomposition of silicon oxynitride salts with the lithium cations occurs in the direction of two compounds: one of a purely oxide coordination surrounding, i.e. lithium silicate, and the second one of a purely nitride surrounding of the coordination center (lithium nitride silicates). Thus, the same types of products of the thermal decomposition of compounds of a mixed oxynitride ligand surrounding were obtained as those described in the previous chapters among the boron and aluminum compounds. The numbers of the reactions correspond to those of the transformations in Fig. 15.

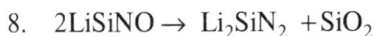

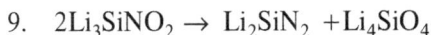

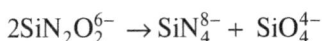

8. $2LiSiNO \rightarrow Li_2SiN_2 + SiO_2$

$2SiNO^- \rightarrow SiN_2^{2-} + SiO_2$

9. $2Li_3SiNO_2 \rightarrow Li_2SiN_2 + Li_4SiO_4$

$2SiNO_2^{3-} \rightarrow SiN_2^{2-} + SiO_4^{4-}$

10. $4Li_5SiNO_3 \rightarrow Li_8SiN_4 + 3Li_4SiO_4$

$4SiNO_3^{5-} \rightarrow SiN_4^{8-} + 3SiO_4^{4-}$

11. $2Li_6SiN_2O \rightarrow Li_8SiN_4 + Li_4SiO_4$

$2SiN_2O_2^{6-} \rightarrow SiN_4^{8-} + SiO_4^{4-}$

3.2.3. Nitride and oxynitride compounds of phosphorus

Phosphorus forms three nitrides of a different composition: P_3N_5, P_4N_6 and PN. Compounds in which PN_2^- and PN_3^{4-} species occur both in salts with lithium cations and in the protonated form and PN_4^{7-} anions in lithium salts are also known. However, PN^- and PN^{3-} species occur exclusively in the protonated form.

In Fig. 16. the classification table with marked nitride compounds of phosphorus is presented.

3.2.3.1. Properties of phosphorus nitride

Phosphorus nitride: P_3N_5 occurs, taking into account all the literature reports, in three crystalline forms: hexagonal α, tetragonal β, and γ. Structure α has lattice constant a = ll.88Å and c = 27, 37Å, and the β form a = 6.063Å and c = 6.822Å. Phosphorus nitride undergoes thermal decomposition at above 700°C transforming to PN with nitrogen evolution:

1. $P_3N_5 \rightarrow 3PN + N_2$

At higher temperatures further decomposition proceeds (the numbers of the reactions correspond to those of the transformations in Fig. 16.):

2. $2PN \rightarrow P_2 + N_2$

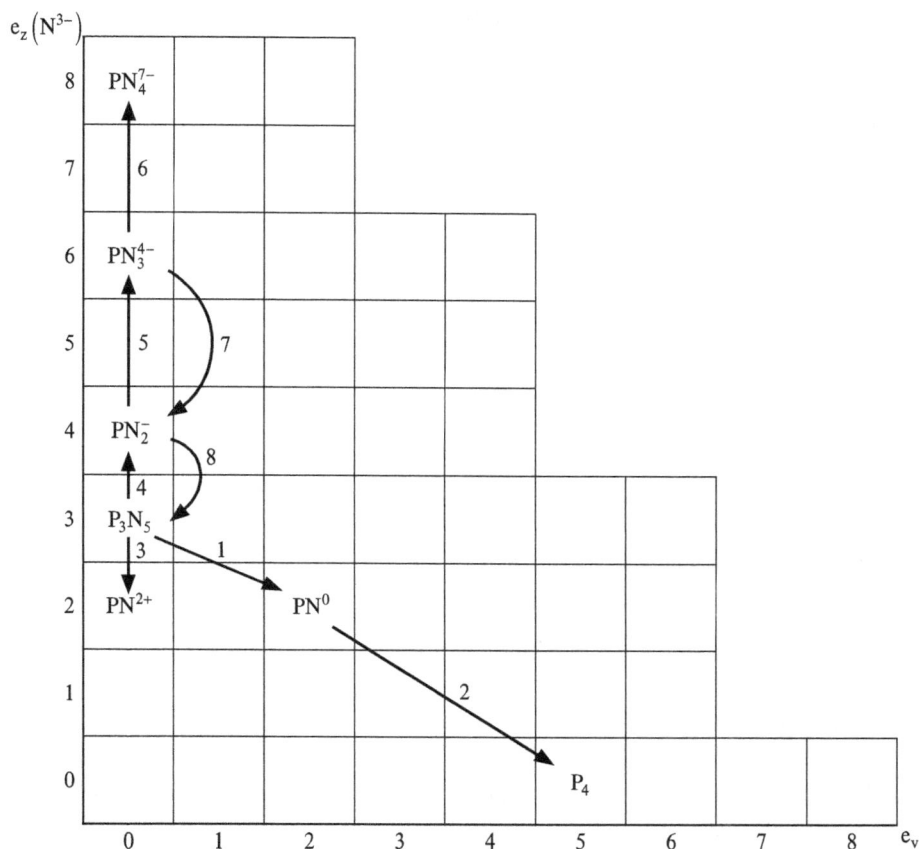

Fig. 16. Nitride compounds of phosphorus

The latter reaction implies the reducing properties of phosphorus nitride at high temperatures. The nitride oxidizes in the air above 700°C according to the reaction:

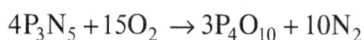

$$4P_3N_5 + 15O_2 \rightarrow 3P_4O_{10} + 10N_2$$

At 700°C it reacts with chlorine

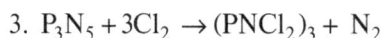

3. $P_3N_5 + 3Cl_2 \rightarrow (PNCl_2)_3 + N_2$

Phosphorus nitride practically does not hydrolyze in water and does not decompose under the influence of hydroxide and acid solutions. Concentrated sulfuric acid, when hot, decomposes P_3N_5 to $(NH_4)_2SO_4$ and H_3PO_4, which is taken advantage of for the analysis of the nitride composition. P_3N_5 reacts with lithium nitride forming complex nitrides (the numbers of the reactions correspond to those of the transformations in Fig. 16):

4. $P_3N_5 + Li_3N \rightarrow 3LiPN_2$

$$P_3N_5 + N^{3-} \rightarrow 3PN_2^-$$

5. $LiPN_2 + Li_3N \rightarrow Li_4PN_3$

$$PN_2^- + N^{3-} \rightarrow PN_3^{4-}$$

6. $Li_4PN_3 + Li_3N \rightarrow Li_7PN_4$

$$PN_3^{4-} + N^{3-} \rightarrow PN_4^{7-}$$

3.2.3.2. Methods of phosphorus nitride preparation

Phosphorus nitride P_3N_5 can be obtained by the following methods:

1. Direct synthesis from elements

$$3P_2 + 10N_2 \rightarrow 2P_3N_5$$

The reaction is carried out at 2250°C under a pressure of 1 at, and then the reactor is rapidly cooled to under the temperature of P_3N_5 decomposition. A similar reaction occurs during quiet electrical discharges (ca. 10 kV) under a pressure of 0.1 at. These methods permit carrying out the process at various temperatures and pressures.

2. Nitriding of phosphorus compounds with ammonia

The reaction can be carried out in the gaseous phase, when the flux of PCl_5 vapors in the carrier gas and NH_3 flux meet in the hot reactor in the 700°C zone.

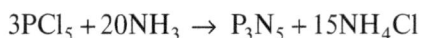

$$3PCl_5 + 20NH_3 \rightarrow P_3N_5 + 15NH_4Cl$$

PCl_5 reacts also with ammonia at low temperatures:

$$PCl_5 + 8NH_3 \rightarrow P(NH)_2NH_2 + 5NH_4Cl$$

Phosphorus amidoimide is formed in this reaction, the protonless skeleton of which is PN_3^{4-}. The thermal decomposition of that compound leads to phosphame and next to nitride (the numbers of the reactions correspond to those of the transformations in the table in Fig. 16.):

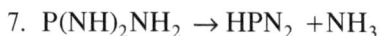

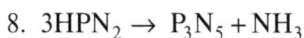

7. $P(NH)_2NH_2 \rightarrow HPN_2 + NH_3$

$$PN_3^{4-} \rightarrow PN_2^- + N^{3-}$$

8. $3HPN_2 \rightarrow P_3N_5 + NH_3$

$$3PN_2^- \rightarrow P_3N_5 + N^{3-}$$

Phosphorus pentasulfide in reaction with ammonia yields phosphorus nitride:

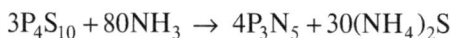

$$3P_4S_{10} + 80NH_3 \rightarrow 4P_3N_5 + 30(NH_4)_2S$$

$NP(SNH_4)_2$, $NPSNH_4SH$, $NP(SH)_2$ and PNS are the intermediates of that process. Phosphorus sulfonitride in reaction with ammonia transforms into nitride:

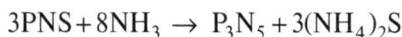

$$3PNS + 8NH_3 \rightarrow P_3N_5 + 3(NH_4)_2S$$

Also phosphorus oxynitride or phosphoryl triamide, in an ammonia flux, transforms into phosphorus nitride:

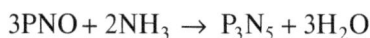

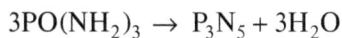

$$3PNO + 2NH_3 \rightarrow P_3N_5 + 3H_2O$$

$$3PO(NH_2)_3 \rightarrow P_3N_5 + 3H_2O$$

The two latter reactions involve the transformation of oxynitride phosphorus compounds, which will be presented in more detail below.

3.2.3.3. Oxynitride compounds of phosphorus

Phosphorus is a successive element lying in the same period as aluminum and silicon, forming compounds of mixed oxynitride coordination. It should be expected that also this element would form oxynitride salts. A whole number of compounds is known of a mixed surrounding of the coordination center: phosphorus oxynitride PNO and derivatives of phosphorus triamide $PO(NH_2)_3$, in which part of the protons are replaced by alkali metal cations. Many glassy phases were obtained based on alkali metal phosphates and nitrides, in which the existence of a mixed ligand surrounding round phosphorus was found [16]. In Fig. 17. the classification table with the marked oxide, nitride and oxynitride species of phosphorus is presented. The routes of preparing phosphorus oxynitride from phosphoryl triamide and phosphorus nitride from phosphoryl triamide or phosphorus oxynitride are marked. Salts of phosphorus oxy acids with alkali metal, beryllium magnesium and alkaline earth metal cations are described in monographs and their properties are well known [17]. On the other hand, nitridephosphates, lying on the $e_z(O^{2-}) = 0$ line, with cations of same alkali metals and beryllium, magnesium and alkaline earth metals have been obtained only in the past forty years. The known oxynitride species (PNO – phosphorus oxynitride, PN_3O^{6-} – protonless skeleton of phosphoryl triamide) and not obtained till now species of a mixed ligand surrounding of the coordination center are shown in the respective fields of the table (the unknown compounds in parentheses). New compounds – phosphorus oxynitride salts – can be obtained in the hypothetic reactions presented below (the numbers of the reactions correspond to those of the transformations in Fig. 17.):

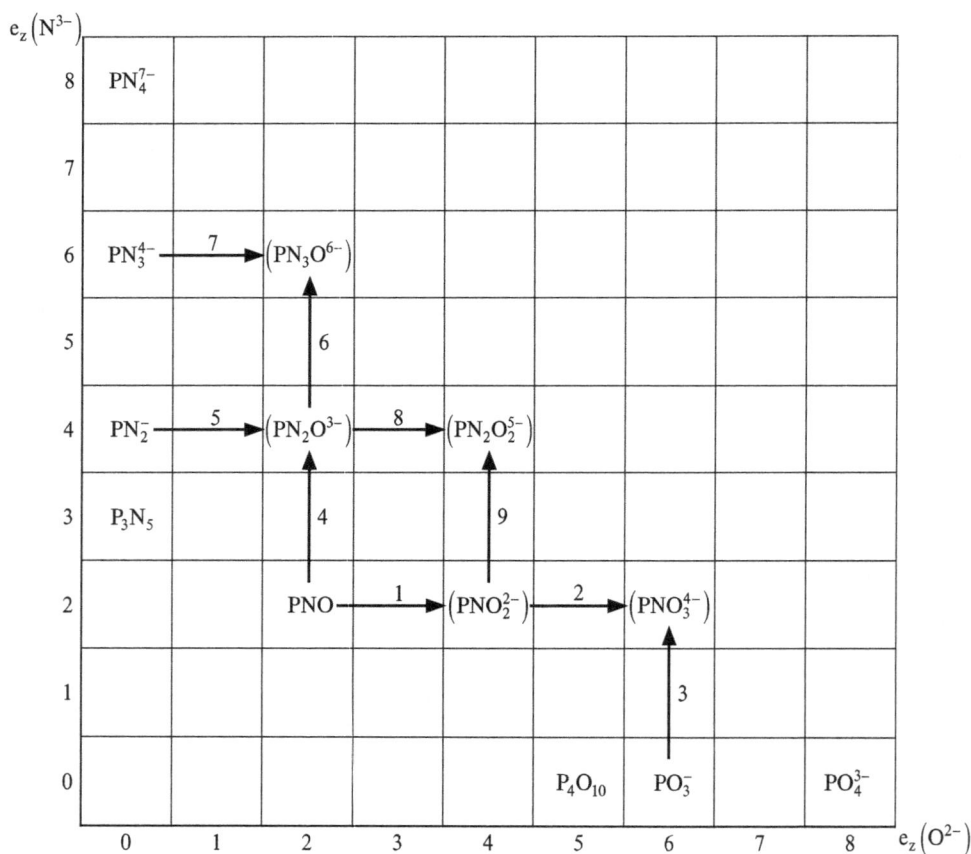

Fig. 17. Oxynitride compounds of phosphorus

1. $PNO + Li_2O \rightarrow Li_2PNO_2$

   $PNO + O^{2-} \rightarrow PNO_2^{2-}$

2. $Li_2PNO_2 + Li_2O \rightarrow Li_4PNO_3$

   $PNO_2^{3-} + O^{2-} \rightarrow PNO_3^{4-}$

3. $LiPO_3 + Li_3N \rightarrow Li_4PNO_3$

   $PO_3^- + N^{3-} \rightarrow PNO_3^{4-}$

4. $PNO + Li_3N \rightarrow Li_3PN_2O$

   $PNO + N^{3-} \rightarrow PN_2O^{3-}$

5. $LiPN_2 + Li_2O \rightarrow Li_3PN_2O$

   $PN_2^- + O^{2-} \rightarrow PN_2O^{3-}$

6. $Li_3PN_2O + Li_3N \rightarrow Li_6PN_3O$

   $PN_2O^{3-} + N^{3-} \rightarrow PN_3O^{6-}$

7. $Li_4PN_3 + Li_2O \rightarrow Li_6PN_3O$

   $PN_3^{4-} + O^{2-} \rightarrow PN_3O^{6-}$

8. $Li_3PN_2O + Li_2O \rightarrow Li_5PN_2O_2$

   $PN_2O^{3-} + O^{2-} \rightarrow PN_2O_2^{5-}$

9. $Li_2PNO_2 + Li_3N \rightarrow Li_5PN_2O_2$

   $PNO_2^{2-} + N^{3-} \rightarrow PN_2O_2^{5-}$

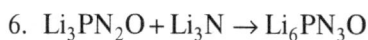

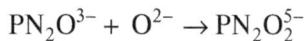

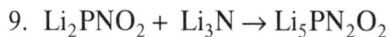

The eventual confirmation by the experimental of the predicted course of processes, when from two different pairs of reactants the same new, unknown till now compound is obtained, would undoubtedly be further proof for the usefulness of the morphological classification for the search of the existence of hypothetic species. As was described in the chapter devoted to the oxynitride compounds of silicon, a similar procedure made it possible to obtain and identify a number of oxynitride-silicates. The search for new phosphorus compounds was also based on the application of a thermal analysis, X-ray phase analysis and classical qualitative and quantitative analysis. It appeared during studies on oxynitride-phosphates that the reactions proposed above indeed occur, which is connected with exothermic effects and the disappearance of substrates, which was confirmed by the X-ray phase analysis, but in the majority of processes amorphous products are formed. Therefore, additional reactions were carried out in which polycentric phosphorus compounds were used: P_4O_{10} and P_3N_5. A classification analysis of the possible reaction courses involving species having more than one coordination center implies some difficulties. However, it enables the explanation of some problems that have not been solved during studies on the reactions of monocentric species. The reactions of phosphorus pentoxide with lithium and magnesium nitrides and of phosphorus nitride with lithium and sodium oxides were studied. The experiments carried out covered the following reactions: lithium oxide with phosphorus nitride, lithium nitride with phosphorus pentoxide, lithium oxide with phosphorus oxynitride, lithium nitride with phosphorus oxynitride, lithium metaphosphate with lithium nitride and magnesium nitride with phosphorus pentoxide and oxynitride. Studies were carried out at different reactants mole ratios. Initial information on the temperatures and courses of the reactions were obtained by thermal analysis methods. The products of the respective processes were identified by an X-ray phase analysis through a comparison with standards. A classical qualitative and quantitative analysis was also applied. The formation of a number of crystalline phases was found, which did not include any of the known compounds of the elements considered. The reaction of lithium oxide with phosphorus nitride at the 1:1 molar ratio at 350°C leads to a phase unknown till now, and at 900°C to a successive new crystalline phase. The reaction of Li_2O with P_3N_5 at the molar ratio 2:1 yields the third crystalline product in the Li-P-N-O system. The same phase is obtained in the reaction of lithium nitride with phosphorus oxynitride (PNO) at 845°C, i.e. from completely different reactants. The following notion is one of the possible schemes of such a reaction:

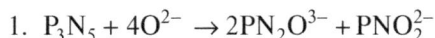

1. $P_3N_5 + 4O^{2-} \rightarrow 2PN_2O^{3-} + PNO_2^{2-}$

The following transformations are another potential route:

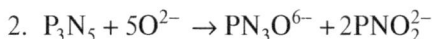

2. $P_3N_5 + 5O^{2-} \rightarrow PN_3O^{6-} + 2PNO_2^{2-}$

or

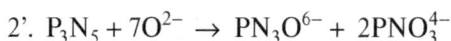

2'. $P_3N_5 + 7O^{2-} \rightarrow PN_3O^{6-} + 2PNO_3^{4-}$

The numbers of the reactions correspond to those of the transformations in the classification table in Fig. 18.

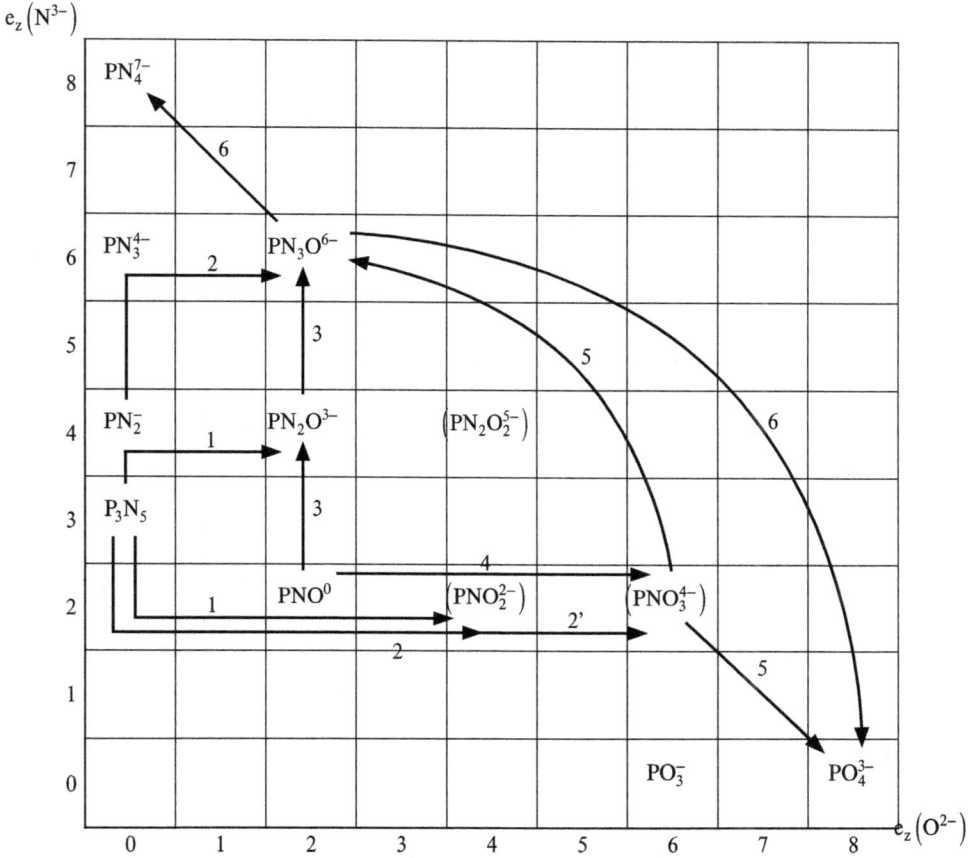

Fig. 18. Oxynitride compounds of phosphorus

As was mentioned earlier, the reaction of phosphorus oxynitride with lithium nitride leads to a crystalline phase identical to that obtained in the reaction of lithium oxide with phosphorus nitride. It is also known that the reaction of phosphorus oxynitride with lithium nitride can proceed only according to the following scheme:

3. $PNO + N^{3-} \rightarrow PN_2O^{3-}$

$$PN_2O^{3-} + N^{3-} \rightarrow PN_3O^{6-}$$

Fields with PN_2O^{3-} and PN_3O^{6-} are the ones where the three routes presented above cross each other. The considerations are complicated by the fact that the same crystalline phase is obtained also in the reaction of phosphorus oxynitride with lithium oxide (at an excess of the latter).

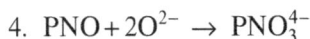

4. $PNO + 2O^{2-} \rightarrow PNO_3^{4-}$

The indication of the transformation route of salt with a PNO_3^{4-} anion to a salt containing a PN_3O^{6-} anion requires the assumption that the first of these compounds would undergo an acidic-basic disproportionation according to the reaction:

5. $3PNO_3^{4-} \rightarrow PN_3O^{6-} + 2PO_4^{3-}$

This reaction course is probable, since Li_3PO_4 was identified as one of the products and could originate from reaction 5. The reaction course of $LiPO_3$ with Li_3N in which the orthophosphate is obtained, instead of the expected hypothetical Li_4PNO_3, is additional confirmation of the instability of the latter salt. The second decomposition product is amorphous, but it could be Li_6PN_3O. This hypothesis is probable in so far as in the works on the reactivity of magnesium nitride with phosphorus oxynitride and phosphorus pentoxide a salt of Mg_3PN_3O stoichiometry was obtained. Thus the obtained lithium and magnesium salts with the PN_3O^{6-} anion would be hexasubstituted derivatives of phosphoryl triamide $PO(NH_2)_3$. An answer to the question as to which of the other hypothetic species placed on the classification table corresponds to the crystalline phases obtained in the presented studies is at present hazardous. Since the physicochemical properties of these phases are completely unknown, it is difficult to obtain them without impurities and in a further stage – crystallization. The present results on the studies of the possibility of obtaining the phosphorus salts with mixed oxynitride coordination surrounding, and especially of their comparison with the classification predictions, confirm the usefulness of the morphological classification for the theoretical preparation of the experiment. The necessity of carrying out corrections and putting forward additional hypotheses during studies does not diminish the advantages of the classification. The course of the thermal decomposition of lithium and magnesium salts with the PN_3O^{6-} anion is in agreement with the general principles formulated earlier and proceeds in the direction of two compounds of purely coordination surrounding:

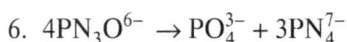

6. $4PN_3O^{6-} \rightarrow PO_4^{3-} + 3PN_4^{7-}$

(process 6 in Fig. 18.)

Identical routes of the thermal decomposition of oxynitride salts were found earlier among boron, silicon and aluminum compounds, which was described in the previous chapters.

3.2.4. Sulfur nitrides

Sulfur forms with nitrogen compounds of a zero charge balance of species of various compositions. There are nitrides: S_4N_4, S_2N_2, SN, S_4N_2 and $S_{11}N_2$. A small difference in electronegativity leads to the occurrence of covalent bonds with the simultaneous occurrence of the capability to form polymers in these compounds, characteristic for sulfur. Fig. 19. presents the classification table with the marked nitride compounds of sulfur. Nitrides S_4N_4, S_2N_2 and SN lie in the same field of the table. However, the nitrides S_4N_2 and $S_{11}N_2$ have fraction values of the e_z numbers, and thus they lie between the fields and have not been included in the table.

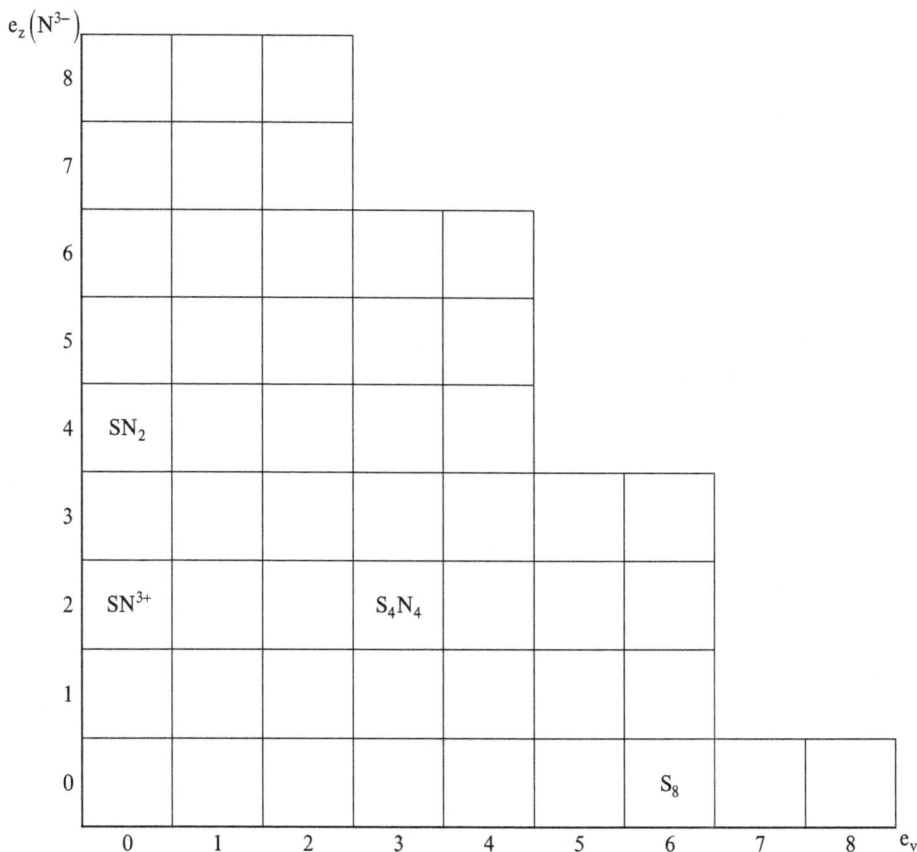

Fig. 19. Nitride compounds of sulfur

Nitrides and also species of a mixed ligand surrounding of the coordination center, e.g. fluoronitride and oxynitride are known. Not all of these compounds can be presented in the table of the $e_z(O^{2-}) - e_z(N^{3-})$ axes. This results from the fact that these are polycentric species.

S_4N_4 is formed in the reaction of liquid ammonia with free sulfur.

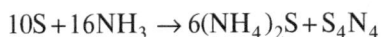

$$10S + 16NH_3 \rightarrow 6(NH_4)_2S + S_4N_4$$

S_4N_4 melts at 178°C, but it can undergo violent decomposition with the evolution of a considerable amount of heat. The sublimation of S_4N_4 under reduced pressure leads, through depolimerization, to the formation of S_2N_2, unstable at normal temperatures. This crystalline substance undergoes transformation above 0°C with the formation of S_4N_4 and SN. This latter nitride is a chain polymer of the electric conduction value characteristic for materials placed between metals and semiconductors. S_4N_2 is formed during the decomposition of S_4N_4 under vacuum or in a solution of carbon disulfide:

$$S_4N_4 \rightarrow S_4N_2 + N_2$$

S_4N_2 is unstable and decomposes with an explosion to free sulfur and nitrogen.

3.3. Nitride and oxynitride compounds of elements of the fourth period

3.3.1. Gallium nitrides

Gallium forms a nitride of the GaN composition formula, a compound of a chemical stability comparable to that of aluminum nitride. The complex lithium-gallium nitride Li_3GaN_2 is known. In Fig. 20. the classification table in the $e_v - e_z$ coordinate system with the marked nitride species of gallium is presented.

3.3.1.1. Properties of gallium nitride

Gallium nitride occurs in a hexagonal structure of the wurtzite type of lattice constants a = 3.189Å and c = 5.186Å. Gallium nitride is characterized by physical properties typical for semiconductors of elements between the third and fifth group of the periodic system. However, it has a much higher melting point (1500°C) than the other compounds of this type: arsenides or antimonides. Also the resistance towards oxidation and reactive media of the gallium nitride is greater than that of the other $A^{III}B^V$ compounds. The data on the vapor pressure over gallium nitride at high temperatures are inconsistent. At 1100°C it is considered as 10^{-11}at. Gallium nitride is a superconductor below 5.18K. As was mentioned earlier, gallium nitride is a chemically stable compound. It does not undergo decomposition in hot solutions of acids, and decomposes slowly in hot solutions of sodium or potassium hydroxides. Fused hydroxides decompose gallium nitride with ammonia evolution:

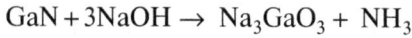

$$GaN + 3NaOH \rightarrow Na_3GaO_3 + NH_3$$

This reaction is a base for determining the content of nitrogen in gallium nitride. The nitride undergoes oxidation above 700°C according to the reaction:

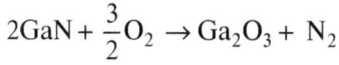

$$2GaN + \frac{3}{2}O_2 \rightarrow Ga_2O_3 + N_2$$

Hydrogen does not react up to 1000°C.

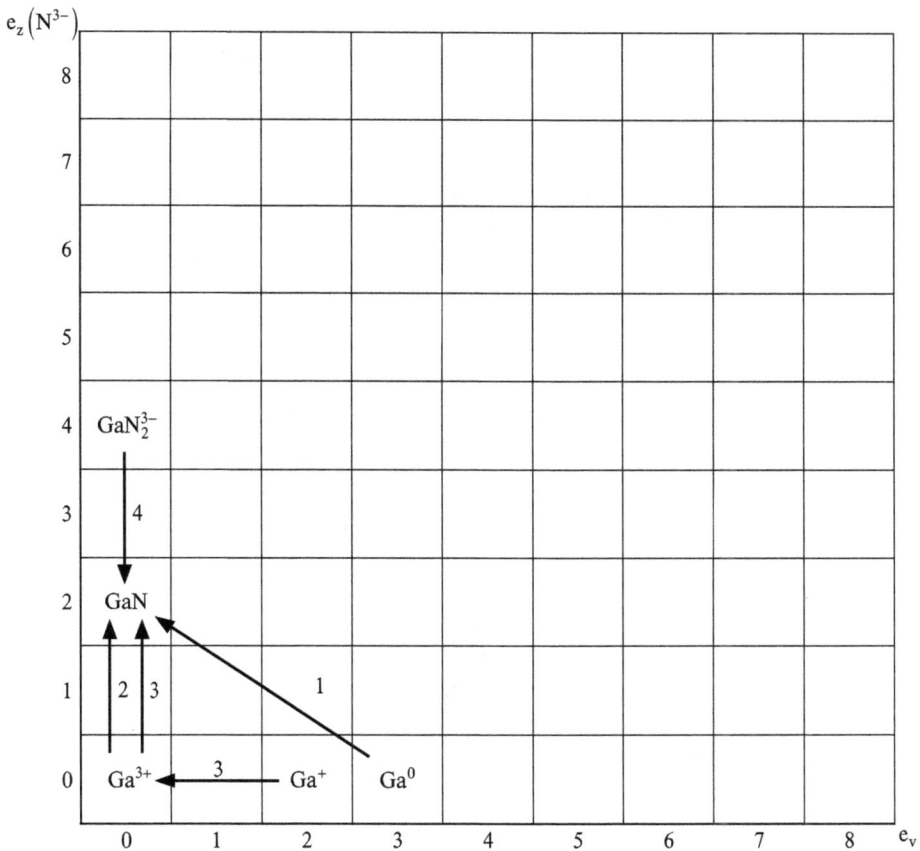

Fig. 20. Nitride compounds of gallium

3.3.1.2. Methods of obtaining gallium nitride

Gallium nitride can be obtained by direct synthesis from elements. The reaction is carried out at 1500°C under a pressure of 5000 at. Practically all the methods of gallium nitride synthesis are based on applying ammonia as the nitriding reagent [18. 19]. This can be gaseous ammonia, in the atmosphere in which the process occurs, i.e. it is introduced to the main reactant (gallium compound) from the outside. "Internal" nitriding takes place in the thermal decomposition of compounds containing nitrogen, adducts with ammonia or ammonium salts. The reaction of metallic gallium with ammonia at 1000–1300°C (transformation 1 in the classification table in Fig. 20.) is the most often applied method of preparing gallium nitride:

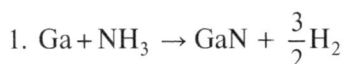

$$1.\ Ga + NH_3 \rightarrow GaN + \frac{3}{2}H_2$$

The main difficulty in carrying out this process is the fact that gallium is a liquid at the reaction temperature (m.p. 29.7°C), and so the interface contact surface is small. The nitride formed covers the metal surface by a compact layer, which strongly limits the diffusion of ammonia. Complete transformation of the metal into the nitride requires the application of additional solutions. Upon cooling, the mixture of gallium and gallium nitride is submitted to grinding, and then the nitriding is repeated. Obtaining a nitride of a composition corresponding to the theoretical one requires the application of lengthy nitriding or multiple grinding. Another modification of the process is adding to gallium a substance which evolves ammonia during thermal decomposition. This causes both "internal" nitriding of the fused metal and mixing it. Ammonium carbonate fulfils these requirements very well. A mixture of gallium with $(NH_4)_2CO_3$ is slowly heated to 1200°C in the ammonia flux. The following reactions proceed under these conditions:

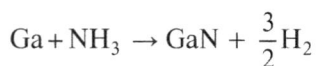

$$(NH_4)_2CO_3 \rightarrow 2NH_3 + H_2O + CO_2$$

$$Ga + NH_3 \rightarrow GaN + \frac{3}{2}H_2$$

The evolution of gaseous materials in large volumes causes violent mixing of the liquid reactants. Above 500°C the mixture solidifies and is still sufficiently porous for nitriding with ammonia introduced from outside to proceed up to 1200°C. However, the nitride obtained in this way does not contain free gallium, but is contaminated by the oxide and impurities from ammonium carbonate, which are difficult to remove. Gallium oxide reacts with metallic gallium forming gallous oxide, volatile at above 700°C, which can be removed from the surface of the metal by evaporation under reduced pressure. From a thus prepared metal a nitride of the smallest contamination with oxygen is obtained in multistage nitriding with ammonia. Temperatures in the

74

1100–1200°C range and high values of the ammonia flow rate over the surface of gallium are recommended in literature reports [18]. Gallium nitride grows in the direct proximity of the metal surface in the form of needles or platelets. Gallium trioxide is a successive reactant used in the synthesis of gallium nitride. In reaction with ammonia it transforms into a nitride with the formation of water, according to the following reaction (transformation 2 in Fig. 20):

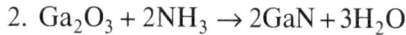

2. $Ga_2O_3 + 2NH_3 \rightarrow 2GaN + 3H_2O$

This method can be applied according to two alternative procedures. The first one consists in nitriding the trioxide in the ammonia flux, i.e. "external" nitriding. The second one consists in preparing a mixture of gallium trioxide and ammonium carbonate and then in a rising temperature in the atmosphere of ammonia. The role of ammonium carbonate is similar as in methods originated from metallic gallium, but the synthesis is based on the reaction of the oxide with ammonia. A number of industrial methods of the synthesis of gallium nitride from the gaseous phase have been established, where the high volatility of the galleous oxide or galleous halides at elevated temperatures is taken advantage of. These processes are based on the following reactions (route 3 in Fig. 20.):

3. $Ga_2O + 2NH_3 \rightarrow 2GaN + H_2O + 2H_2$

$GaCl + 2NH_3 \rightarrow GaN + NH_4Cl + H_2$

The reactions carried out in the section of the reactor characterized by lower temperatures are the source of the oxide or galleous chloride:

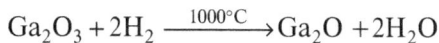

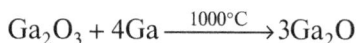

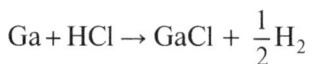

$$Ga_2O_3 + 2H_2 \xrightarrow{1000°C} Ga_2O + 2H_2O$$

$$Ga_2O_3 + 4Ga \xrightarrow{1000°C} 3Ga_2O$$

$$Ga + HCl \rightarrow GaCl + \frac{1}{2}H_2$$

The vapors of these compounds are introduced to the reaction zone (1100°C) in the flux of the carrier gas, where they meet with ammonia. These methods make it possible to deposit gallium nitride on prepared bases – including depositing large dimensions. The synthesis of gallium nitride from the reaction of gallium trichloride with ammonia in the gaseous phase is also possible (route 2 in Fig. 20.):

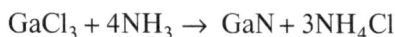

$$GaCl_3 + 4NH_3 \rightarrow GaN + 3NH_4Cl$$

The course of the thermal decomposition of $GaBr_3 \cdot 4NH_3$ in the atmosphere of ammonia should be placed similarly in the classification table:

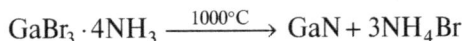

$$GaBr_3 \cdot 4NH_3 \xrightarrow{\quad 1000°C \quad} GaN + 3NH_4Br$$

Also the thermal decomposition of ammonium hexafluorogalate in the ammonia flux leads to gallium nitride:

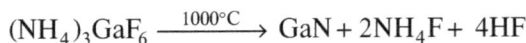

$$(NH_4)_3GaF_6 \xrightarrow{\quad 1000°C \quad} GaN + 2NH_4F + 4HF$$

Trimethylgallium in reaction with ammonia yields the nitride:

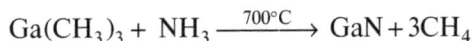

$$Ga(CH_3)_3 + NH_3 \xrightarrow{\quad 700°C \quad} GaN + 3CH_4$$

In this case the product is contaminated by polymers containing gallium, carbon, nitrogen and hydrogen.

The decomposition of LI_3GaN_2 leads to gallium nitride according to the following reaction (process 4 in the classification table in Fig. 20.):

In this case the product is contaminated by polymers containing gallium, carbon, nitrogen and hydrogen. The decomposition of LI_3GaN_2 leads to gallium nitride according to the following reaction (process 4 in the classification table in Fig. 20.):

$$4.\ Li_3GaN_2 \xrightarrow{\quad 600°C \quad} GaN + Li_3N$$

Both gallium phosphide and arsenide transform under the influence of ammonia into nitride:

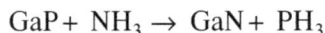

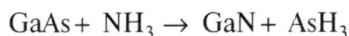

$$GaP + NH_3 \rightarrow GaN + PH_3$$

$$GaAs + NH_3 \rightarrow GaN + AsH_3$$

3.3.2. Nitride and oxynitride compounds of germanium

Germanium forms a considerable number of nitride compounds with a coordination surrounding occurring in the form of salts or as compounds with hydrogen. In Fig. 21. the classification table of germanium species with nitride ligands is presented.

The species $Ge_2N_3^-$, GeN_2^{2-}, GeN_3^{5-}, and GeN_4^{8-} can be isolated as fragments of the anionic sublattice in salts with cations of lithium or beryllium, magnesium, alkaline earth metals and also in compounds with hydrogen – amides or imides [1, 2].

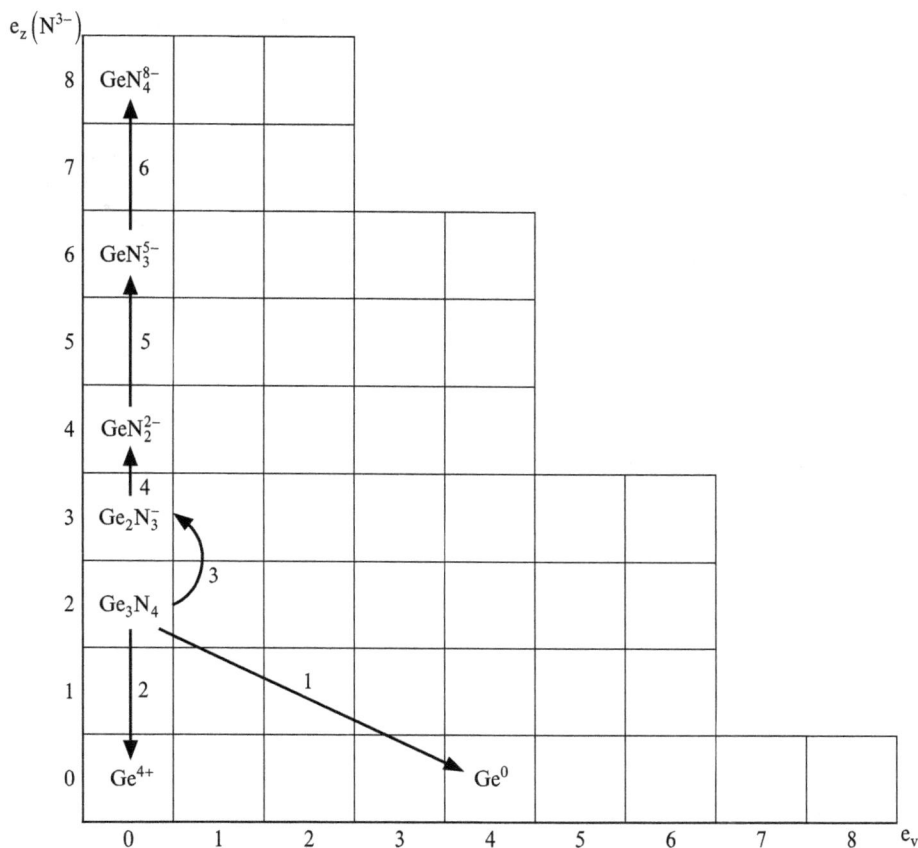

Fig. 21. Nitride compounds of germanium

3.3.2.1. Properties of germanium nitride

Germanium nitride Ge_3N_4 occurs in the crystallographic forms: α and β. The α form is of a rhombic structure of lattice constant a = 4.10Å; b = 7.10Å; c = 5.94Å, and the β form is of a rhombohedral one (a = 8.62Å and α = 108°). Germanium nitride is a thermally and chemically less stable compound in comparison with nitrides of elements lying in the neighborhood of germanium in the periodic system – silicon and gallium. Ge_3N_4 undergoes thermal decomposition to free elements in an atmosphere of nitrogen at 850°C according to the reaction:

1. $Ge_3N_4 \rightarrow 3Ge + 2N_2$

Germanium nitride undergoes oxidation in the air at 750°C:

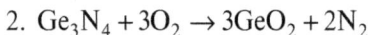

2. $Ge_3N_4 + 3O_2 \rightarrow 3GeO_2 + 2N_2$

77

As far as the thermal decomposition can be presented in the classification table in the $e_v - e_z$ coordinate system (transformation 1 in Fig. 21.), the oxidation of germanium nitride can be shown in the table of axes $e_z(O^{2-}) - e_z(N^{3-})$. An exchange of nitride ligands for oxide ones takes place here. This process can be expressed only formally as the deanionization of the nitride to the Ge^{4+} cation (route 2 in Fig. 21.). Germanium nitride reacts with strong donors of nitride anions – ionic nitrides of lithium, beryllium, magnesium and alkaline earth metals with the formation of nitridegermanates. The synthesis of these compounds is presented below on the example of lithium compounds (the numbers of the reactions correspond to those of the transformations in Fig. 21.):

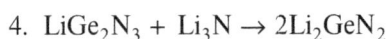

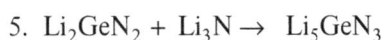

3. $2Ge_3N_4 + Li_3N \rightarrow 3LiGe_2N_3$

$2Ge_3N_4 + N^{3-} \rightarrow 3Ge_2N_3^-$

4. $LiGe_2N_3 + Li_3N \rightarrow 2Li_2GeN_2$

$Ge_2N_3^- + N^{3-} \rightarrow 2GeN_2^{2-}$

5. $Li_2GeN_2 + Li_3N \rightarrow Li_5GeN_3$

$GeN_2^{2-} + N^{3-} \rightarrow GeN_3^{5-}$

6. $Li_5GeN_3 + Li_3N \rightarrow Li_8GeN_4$

$GeN_3^{5-} + N^{3-} \rightarrow GeN_4^{8-}$

It is not quite certain whether the presented reactions proceed in such an order. The kind of nitridegermanate formed depends on the stoichiometry of the reactant mixture and the reaction pathway has not been fully explained yet. In the case of azosilanes synthesis (which was described in section 3.2.2.2.), compounds of complete coordination surrounding were first formed irrespective of the reactant molar ratio. Subsequently, these compounds underwent synproportionation with not complete coordinates yielding species of intermediate e_z values. It cannot be excluded that a similar phenomenon takes place in the case of complex lithium-germanium nitrides.

3.3.2.2. Methods of germanium nitride preparation

Germanium nitride is obtained in processes where ammonia is usually the nitriding reactant. Elemental nitrogen does not react with germanium and even less with germanium compounds. The reaction of germanium with ammonia is the most often applied method from among the many methods of germanium nitride synthesis. This process has to be carried out below the temperature of Ge_3N_4 decomposition

(850°C). The synthesis is usually carried out between 800–825°C, when the rate of nitriding is the greatest and the decomposition of the products does not yet proceed:

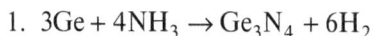

1. $3Ge + 4NH_3 \rightarrow Ge_3N_4 + 6H_2$

The evolution of a considerable amount of hydrogen is a certain inconvenience of this method. The classification table, in which the transformations leading to the formation of germanium nitride are marked, is presented in Fig. 22. The α form of Ge_3N_4 is mainly obtained from the reaction of germanium with ammonia (over 90wt-% in the mixture with β-Ge_3N_4). The time of the synthesis is recommended to be from 2 to 8h. Ammonium carbonate is added to the germanium powder in order to increase the contact surface of the solid and gaseous phases. Ammonium carbonate evolves ammonia during thermal decomposition, which increases the NH_3 pressure inside the sample and also causes partial mixing of the powder:

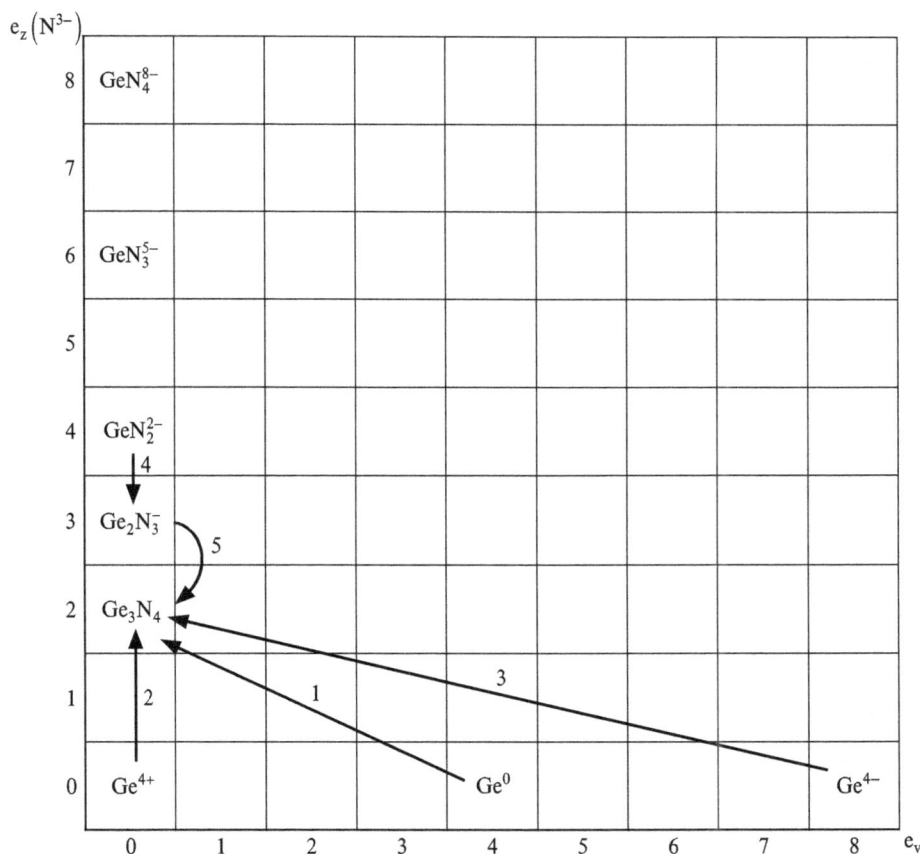

Fig. 22. Nitride compounds of germanium

$$\left(NH_4\right)_2 CO_3 \rightarrow 2NH_3 + H_2O + CO_2$$

The reaction of germanium dioxide with ammonia is the second method of germanium nitride synthesis. This process should also be carried out below the temperature of the nitride decomposition, i.e. up to 825°C:

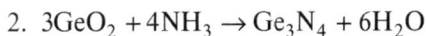

2. $3GeO_2 + 4NH_3 \rightarrow Ge_3N_4 + 6H_2O$

Also in this case ammonium carbonate can be added to germanium dioxide. Similarly as the oxidation of germanium nitride with oxygen, also the formation of nitride from germanium dioxide is fully presented in the classification table in the $e_v - e_z(N^{3-})$ coordinate system. The germane from the reaction of ammonia or hydrazine also forms germanium nitride:

3. $3GeH_4 + 4NH_3 \rightarrow Ge_3N_4 + 12H_2$

This process is applied for depositing thin nitride layers on solid bases. The rise of the Ge_3N_4 layer starts from 300°C and the rate of depositing increases with a rise of temperature up to 800°C. Above that temperature the nitride decomposition begins. The thermal decomposition of germanium imide at 350°C in an ammonia atmosphere leads to germanium nitride:

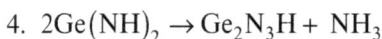

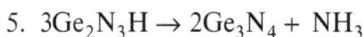

4. $2Ge(NH)_2 \rightarrow Ge_2N_3H + NH_3$

$$2GeN_2^{2-} \rightarrow Ge_2N_3^- + N^{3-}$$

5. $3Ge_2N_3H \rightarrow 2Ge_3N_4 + NH_3$

$$3Ge_2N_3^- \rightarrow 2Ge_3N_4 + N^{3-}$$

These reactions are presented under the respective numbers in the classification table in Fig. 22. It should be noticed that these are deanionization processes contrary to the earlier described synthesis of nitridegermanates from lithium nitride and germanium nitride.

3.3.2.3. Oxynitride compounds of germanium

The oxynitride compounds of boron, carbon, aluminum, silicon and phosphorus described in the previous chapters have been obtained during the last thirty years. The prediction of their existence and ways of preparation was based on the morphological classification of simple species. It appeared that with the finding of hitherto unknown salts of a mixed oxynitride surrounding of the coordination center the usefulness of the morphological classification for predicting the possibility of the existence of preparative methods and the basic chemical properties of successive hypothetic compounds became more clear. The procedure of studies on oxynitride salts is described in the chapter devoted to the silicon species. The results obtained in this way, including the verification of classification predictions, are an encouragement to transfer the evaluated approach to the chemistry of other elements. Theoretical predictions concerning new coordination centers (e.g. germanium) are possible here, as well as considerations of a much wider range, covering the coordination surrounding other than the oxynitride (e.g. fluoronitride).

$e_z(N^{3-})$	0	1	2	3	4	5	6	7	8
8	SiN_4^{8-}								
7									
6	SiN_3^{5-}		(SiN_3O^{7-})						
5									
4	SiN_2^{2-}		(SiN_2O^{4-})		$SiN_2O_2^{6-}$				
3	$Si_2N_3^{-}$								
2	Si_3N_4	Si_2N_2O	$SiNO^{-}$		$SiNO_2^{3-}$		$SiNO_2^{5-}$		
1									
0					SiO_2		SiO_3^{2-}		SiO_4^{4-}

(horizontal axis: $e_z(O^{2-})$)

Fig. 23. Oxynitride compounds of silicon

In Fig. 23. the classification table with the hitherto obtained oxide, nitride and oxynitride monocentric (with some exceptions) species of silicon is presented. In addition, for comparison, a table with the marked germanium species is presented (Fig. 24.).

$e_z(N^{3-})$

GeN_4^{8-} · 8 · · 6 · · · · · ·

7

GeN_3^{5-} · 6 · (GeN_3O^{7-}) · · · · · ·

5 · · 7 · · · · · ·

GeN_2^{2-} · 4 · 5 · (GeN_2O^{4-}) · 5 · $(GeN_2O_2^{6})$ · · · ·

$Ge_2N_3^-$ · 3 · · · 4 · · 6 · · ·

Ge_3N_4 · Ge_2N_2O · $GeNO^-$ · 3 / 2 · $GeNO_2^{3-}$ · $(GeNO_2^{5-})$ · · ·

1 · · · 4 · 1 · 7 · · · ·

0 · · · GeO_2 · GeO_3^{2-} · GeO_4^{4-} · ·

0 · 1 · 2 · 3 · 4 · 5 · 6 · 7 · 8 · $e_z(O^{2-})$

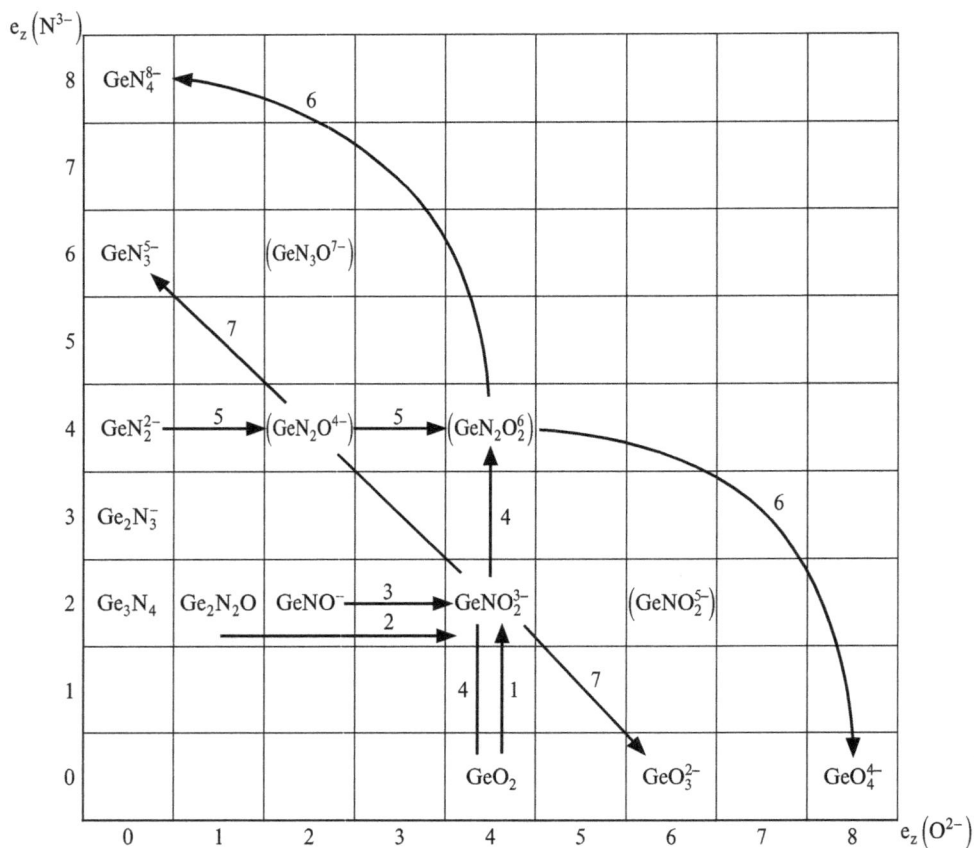

Fig. 24. Oxynitride compounds of germanium

As can be seen, in the range of nitride and oxynitride compounds (Ge_2N_2O – Si_2N_2O, $GeNO^-$ – $SiNO^-$) an exact duplication of structures takes place. In the field of compounds of a mixed oxynitride coordination surrounding, it could be expected that also germanium will form salts containing anions of structures like those obtained with silicon. Thus, the hypothetic oxynitride species of germanium are marked in the classification table in Fig. 24. in parentheses. The salts of alkali metals with $GeNO^-$ anions had already been obtained in the 1980s. Nothing stands in the way of putting forward the thesis that other species can exist: $GeNO_2^{3-}$ (analog of $SiNO_2^{3-}$), $GeNO_3^{5-}$ (analog of $SiNO_3^{5-}$), $GeN_2O_2^{6-}$ (analog of $SiN_2O_2^{6-}$) or others presented in the table. A route of obtaining the hypothetic germanium oxynitride salt Li_3GeNO_2 can also be proposed:

1. $GeO_2 + Li_3N \rightarrow Li_3GeNO_2$

   $GeO_2 + N^{3-} \rightarrow GeNO_2^{3-}$

2. $Ge_2N_2O + 3Li_2O \rightarrow 2Li_3GeNO_2$

   $Ge_2N_2O + 3O^{2-} \rightarrow 2GeNO_2^{3-}$

3. $NaGeNO + Na_2O \rightarrow Na_3GeNO_2$

   $GeNO^- + O^{2-} \rightarrow GeNO_2^{3-}$

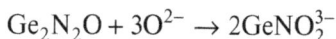

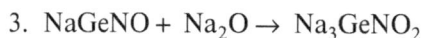

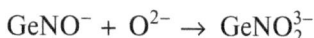

The numbers of the reactions correspond to those of the transformations in the table in Fig. 24. The reaction routes leading to successive hypothetic compounds were predicted according to a similar scheme:

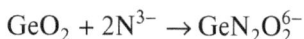

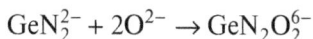

4. $GeO_2 + 2Li_3N \rightarrow Li_6GeN_2O_2$

   $GeO_2 + 2N^{3-} \rightarrow GeN_2O_2^{6-}$

5. $Li_2GeN_2 + 2LI_2O \rightarrow Li_6GeN_2O_2$

   $GeN_2^{2-} + 2O^{2-} \rightarrow GeN_2O_2^{6-}$

A much greater number of predicted reactions can be presented – as many as there are elementary steps in the table connecting the particular structures. However, this is not necessary, since all these processes are simple acts of adding oxide or nitride to species without complete coordination surrounding, and proceeding at a constant value of one of the two e_z numbers. Attempts to predict the thermal decomposition routes of hypothetic salts with a mixed oxynitride surrounding of germanium are also possible. According to the regularities observed earlier in the chemistry of boron, carbon, silicon and phosphorus, these processes should proceed in the direction of two products: one of a purely oxide coordination surrounding and the second of a purely nitride one round germanium.

6. $2Li_6GeN_2O_2 \rightarrow Li_8GeN_4 + Li_4GeO_4$

   $2GeN_2O_2^{6-} \rightarrow GeN_4^{8-} + GeO_4^{4-}$

The reliability of the hypothesis presented was verified by experimental results.

The reaction course of germanium dioxide with lithium nitride and of germanium oxynitride with lithium oxide was studied by means of thermal analysis, X-ray phase analysis and classical chemical analysis. It appeared that in both reactions the same product is formed: Li_3GeNO_3. This result is in agreement with the predictions presented earlier and described by reactions 1 and 2 (transformations 1 and 2 in Fig. 24. correspond to the following):

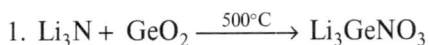

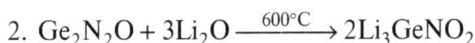

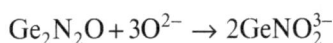

1. $Li_3N + GeO_2 \xrightarrow{\text{500°C}} Li_3GeNO_3$

$GeO_2 + N^{3-} \rightarrow GeNO_2^{3-}$

2. $Ge_2N_2O + 3Li_2O \xrightarrow{\text{600°C}} 2Li_3GeNO_2$

$Ge_2N_2O + 3O^{2-} \rightarrow 2GeNO_2^{3-}$

Li_3GeNO_3 is thermally stable up to 1000°C, when it undergoes decomposition to oxygermanate and azogermanate according to the reaction:

7. $3Li_3GeNO_2 \rightarrow 2Li_2GeO_3 + Li_5GeN_3$

$3GeNO_2^{3-} \rightarrow 2GeO_3^{2-} + GeN_3^{5-}$

The course of the processes described is, as far as concerns the directions of transformations and kinds of products, identical to the analogous reactions of silicon compounds. It is fully concordant with the predictions. The structure of the new salt was not determined.

3.3.3. Arsenic nitrides

There are no reports on the existence of stable arsenic compounds with nitrogen, arsenic nitrides or complex nitrides of that element.

3.3.4. Selenium nitrides

Selenium forms two compounds with nitrogen: Se_4N_4 and $(SeN)_n$. The monomeric nitride Se_4N_4 has the structure of a membered ring of the shape and kind of bonds as in S_4N_4. In the crystalline form it appears in the monoclinic type of lattice of the following parameters a = 9.65Å; b = 9.73Å; c = 6.97Å; β = 104.9°. The $(SeN)_n$ polymer is of a monoclinic structure in the crystalline form. Se_4N_4 undergoes a slow decomposition to elements at 220°C. It hydrolyzes in water with ammonia evolution. Se_4N_4 is formed in the following reactions:

$$12SeCl_4 + 64NH_3 \rightarrow 3Se_4N_4 + 48NH_4Cl + 2N_2$$

$$12Se(NH)_2 \rightarrow 3Se_4N_4 + 8NH_3 + 2N_2$$

$$12SeO(OC_2H_5)_2 + 16NH_3 \rightarrow 3Se_4N_4 + 24C_2H_5OH + 12H_2O + 2N_2$$

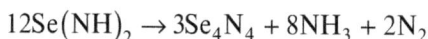

3.4. Nitride compounds of elements of the fifth period

3.4.1. Indium nitride

The number of nitride species of sp elements decreases with a decrease in the electronegativity of those elements. Until now only one such compound of indium was reported – InN. It crystallizes in a hexagonal, fully packed structure of lattice constants equal to a = 3.53Å; c = 5.69Å. There are no reports on the formation of complex indium nitrides with cations of considerably lower electronegativity, such as those known among aluminum or gallium compounds, Li_3AlN_2 or Li_3GaN_2. The hardness and thermal stability of InN are much lower in comparison with aluminum or gallium nitrides. Its chemical stability is also much lower. It undergoes oxidation in the air at 300°C according to the reaction:

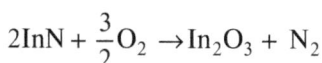

$$2InN + \frac{3}{2}O_2 \rightarrow In_2O_3 + N_2$$

InN decomposes in solutions of hydrochloric, sulfuric and nitric acids, and also of alkali metal hydroxides. Indium nitride is generally obtained in reactions where ammonia is the nitriding agent, resulting from the decomposition of a compound containing indium, or introduced from outside. Only thin layers of the nitride are produced by the deposition of the compound formed from indium vapors under the pressure of nitrogen of ca. 10^{-1} at. during glow discharges. Metallic indium does not undergo any reaction with nitrogen or ammonia at 300–390°C under normal pressure. The thermal decomposition of the polymerized adduct of indium chloride with ammonia leads to indium nitride:

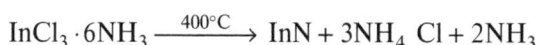

$$InCl_3 \cdot 6NH_3 \xrightarrow{400°C} InN + 3NH_4 Cl + 2NH_3$$

The thermal decomposition of ammonium hexafluoroindium also yields the nitride:

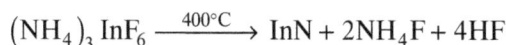

$$(NH_4)_3 InF_6 \xrightarrow{400°C} InN + 2NH_4F + 4HF$$

The nitriding of indium trioxide in an ammonia flux for a few hours at 600°C leads to the nitride:

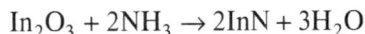

$$In_2O_3 + 2NH_3 \rightarrow 2InN + 3H_2O$$

Ammonium carbonate is added to indium trioxide in order to increase the surface of the solid phase. Ammonia evolves during the thermal decomposition of ammonium carbonate and thus increases the pressure of that compound inside the In_2O_3 powder.

3.4.2. Nitride compounds of tin

In the literature there are reports only on the existence of stannic nitride Sn_3N_4. There are no precise data on its properties and structure. However, it can be presumed that it will be possible to obtain complex stannic nitrides with elements of a considerably lower electronegativity and also of oxynitride salts, such as among germanium compounds.

3.4.3. Antimony nitrides

There are no confirmed data on the formation of stable antimony nitrides.

3.4.4. Tellurium nitrides

A polymeric compound $(Te_4N_4)_n$ is formed in the reaction of tellurium tetrabromide with liquid ammonia. Its structure probably originates from the eight-membered S_4N_4 ring. The physical and chemical properties of that compound have not been described in the literature.

3.5. Nitrides of elements of the sixth period

3.5.1. Thallium nitrides

Thallium, an element of a considerable lower electronegativity in comparison with earlier described elements, forms three compounds with nitrogen: nitrides TlN, Tl_3N and azide TlN_3. In Fig. 25. the classification table with the marked thallium nitrides (the azide is not included in this classification system) is presented. No nitrides of a higher coordination of nitrogen round thallium, i.e. complex nitrides, have been obtained until now.

3.5.2. Lead, bismuth and polon nitrides

There are no confirmed literature reports on the formation of lead, bismuth or polon nitrides.

86

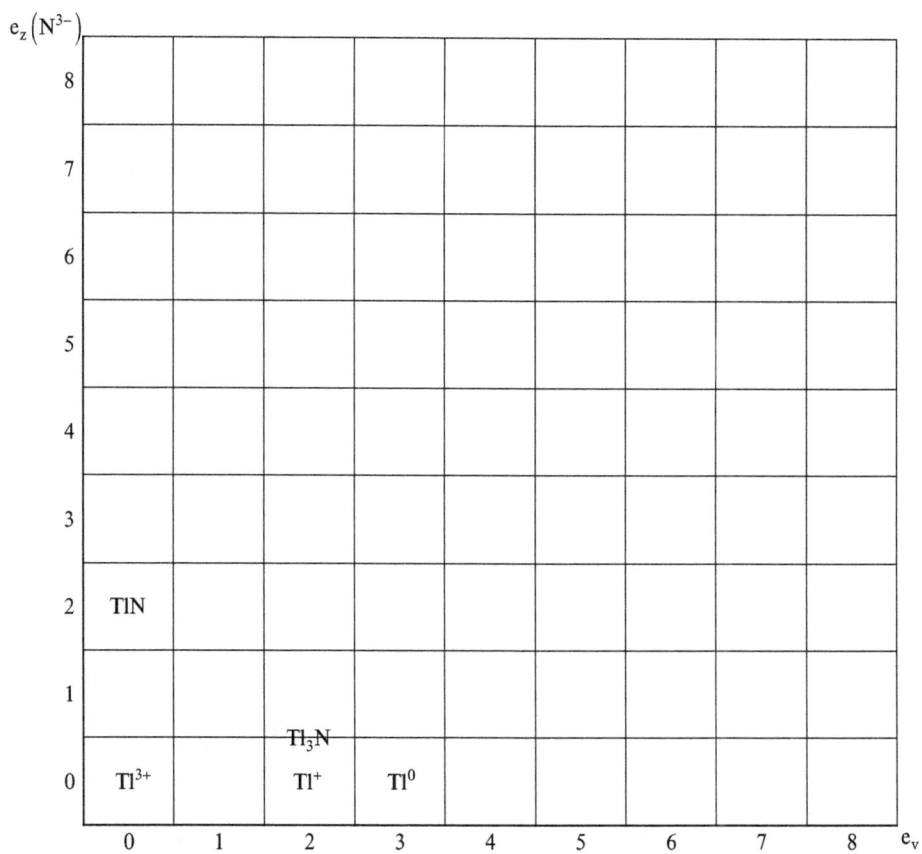

Fig. 25. Nitride compounds of thallium

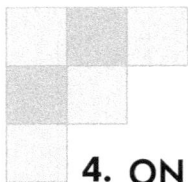

4. ON THE POSSIBILITY OF THE EXISTENCE OF FLUORONITRIDE COMPOUNDS OF THE SP ELEMENTS

In the previous chapter the possibility of applying the morphological classification for compounds of a mixed coordination surrounding other than an oxynitride one was mentioned.

In the course of studies on the reactivity of boron nitride with strong anion donors the reaction with lithium fluoride was carried out. Together with the earlier studies on the anionizing action of lithium oxide and nitride, it gives the possibility to compare the boron tendency to form a mixed oxynitride and fluoronitride coordination surrounding.

The initial studies on obtaining fluoronitride species of boron and silicon do not exclude the existence of a whole series of compounds of this type. The experiments were planned solely on the basis of the morphological classification. In Fig. 26 a table in the $e_z(F^-) - e_z(N^{3-})$ coordinate system is presented, where known fluoride and nitride species of boron and hypothetic fluoronitride species of that element are marked.

The hypothetic fluoronitride anions of boron can be obtained in the following reactions:

1. $BN + F^- \rightarrow BNF^-$

$BNF^- + F^- \rightarrow BNF_2^{2-}$

$BNF_2^{2-} + F^- \rightarrow BNF_3^{3-}$

2. $BF_3 + N^{3-} \rightarrow BNF_3^{3-}$

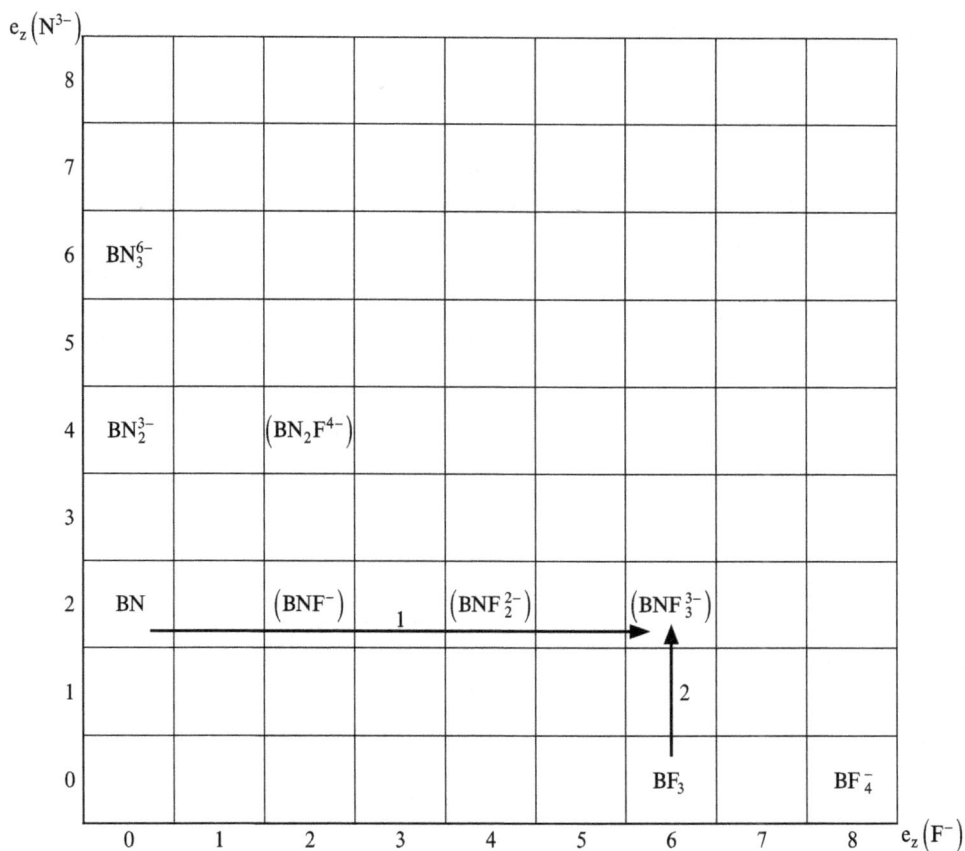

Fig. 26. Fluoronitride compounds of boron

In Fig. 27. a classification table with known fluoride, nitride and hypothetic fluoronitride species of silicon ($SiNF$ is known) is presented.

The eventual routes of obtaining chosen hypothetic fluoronitride species of silicon are presented below:

1. $SiNF + 3F^- \rightarrow SiNF_4^{3-}$

2. $SiF_4 + N^{3-} \rightarrow SiNF_4^{3-}$

3. $SiN_2^{2-} + F^- \rightarrow SiN_2F^{3-}$

4. $SiNF + N^{3-} \rightarrow SiN_2F^{3-}$

5. $SiN_3^{5-} + F^- \rightarrow SiN_3F^{6-}$

6. $SiNF + 2N^{3-} \rightarrow SiN_3F^{6-}$

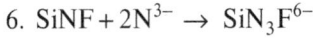

In the presented hypothetic reactions it is possible to obtain the same products from different substrates and in different processes. As was described in the chapters devoted to the oxynitride compounds of silicon and phosphorus, carrying out the experiment in this way considerably simplifies the interpretation of the results in the case of obtaining new phases, hitherto unknown.

The existence of new fluoronitride salts should be expected among aluminum, germanium and phosphorus compounds.

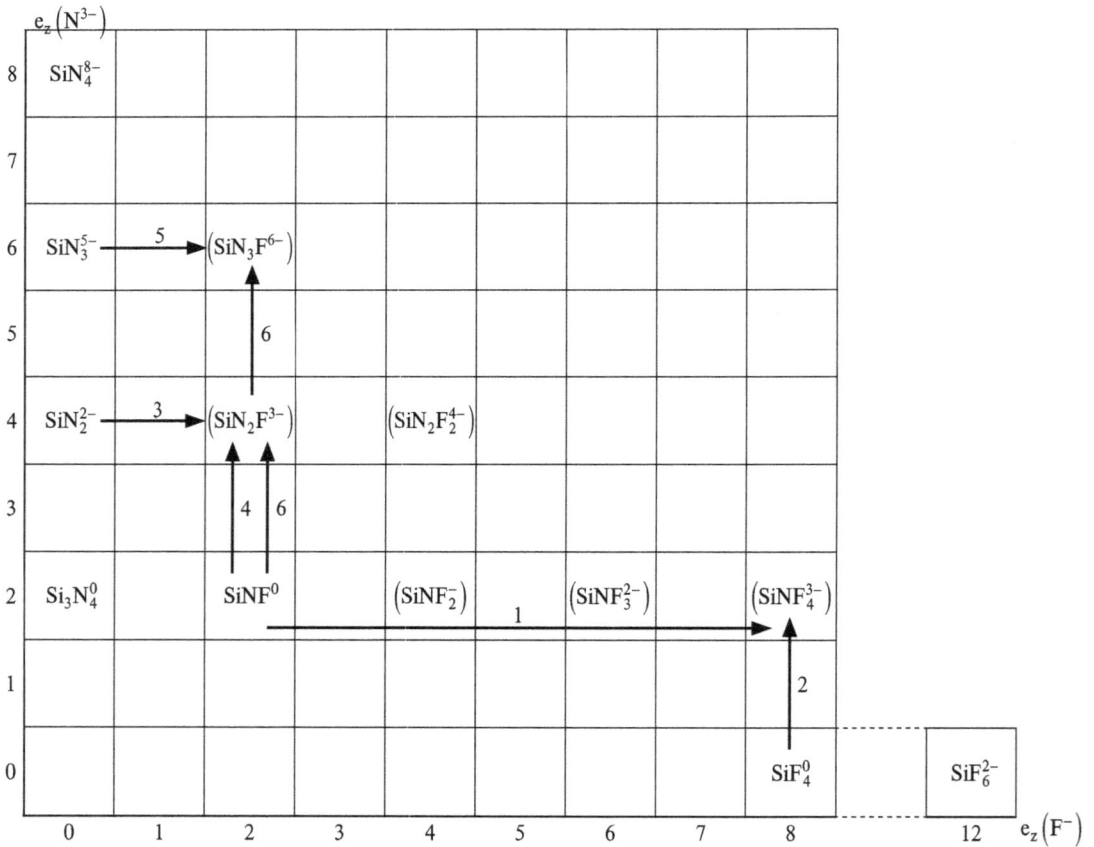

Fig. 27. Fluoronitride compounds of silicon

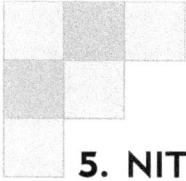

5. NITRIDES OF TRANSITION AND INTRATRANSITION ELEMENTS

Nearly all the transition and intratransition metals occur in one of the four types of crystalline structures: regular, regular face centered, hexagonal and hexagonal close packed. The reactions of these elements with free nitrogen, boron or carbon, occurring below the melting point of the metal, lead to nitrides, borides or carbides. As far as concerns the metal sublattice, the new compound very often has an identical type of crystalline structure as the initial metal. The carbon, boron or nitrogen are inbuilt in the interstitial positions. The four basic types of metal structures with marked sites where nitrogen can be located are presented in Fig. 28. Compounds known as interstitial nitrides, i.e. such in which nitrogen is placed at interstitial positions, are formed when the ratio of nitrogen and metal radii is less than 0.59. This rule is known as the Hägg condition [20–22].

$$R_N : R_M < 0.59 \qquad (r_N : r_M < 0.59)$$

The newly formed interstitial compound can have a different crystalline structure than the initial metal, but usually it is one of the four types of structures presented in Fig. 28. The positions accessible for nitrogen can be occupied in full or in part, hence several types of interstitial nitrides are known. If nitrogen occupies all the interstitial sites, then the nitride has the MN composition. If half of the interstitial sites are occupied, then a compound of the M_2N formula is formed. The occupation of one third of the interstitial sites leads to M_3N. If nitrogen occupies every fourth interstitial site, then M_4N is formed. Among the iron nitrides, Fe_8N is known, in which one of every eight interstitial site is occupied. Such a structure of the nitrides of transition and intratransition elements is directly connected with the physical properties of compounds. The metallic character of bonds causes high (comparable with metals) values of electric and heat conduction and also very good mechanical strength.

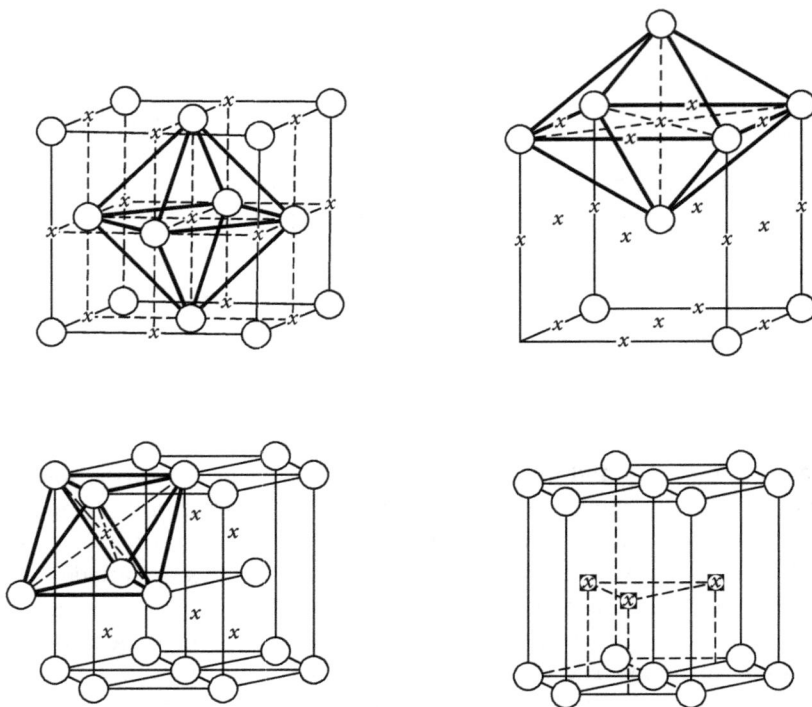

Fig. 28. The structures of interstitial compounds

Also the chemical properties of interstitial nitrides are different than of ionic or covalent compounds. In reaction with donors of a hydrogen cation they exhibit reducing properties:

$$Fe_2N + 4HCl \rightarrow 2FeCl_2 + NH_3 + \frac{1}{2}H_2$$

Nearly all the interstitial nitrides exhibit a "region of homogeneity", i.e. the compound maintains its crystalline structure and does not change its properties, despite changes of the elemental composition varying within a certain range. The homogeneity regions of consecutive nitrides of transition elements are marked in Fig. 29 [1].

The deviations mentioned above indicate a nitrogen deficiency with respect to the theoretical considerations presented earlier. Not all the interstitial sites permitted for a given stoichiometry are occupied in those compounds.

A conventional division of nitrides was presented in the first chapter depending on the type of bonds occurring between the atoms of elements forming this kind of compounds with nitrogen atoms. Transition element nitrides were, according to this division, isolated considering the occurrence of direct metal-metal bonds in the structure of compounds.

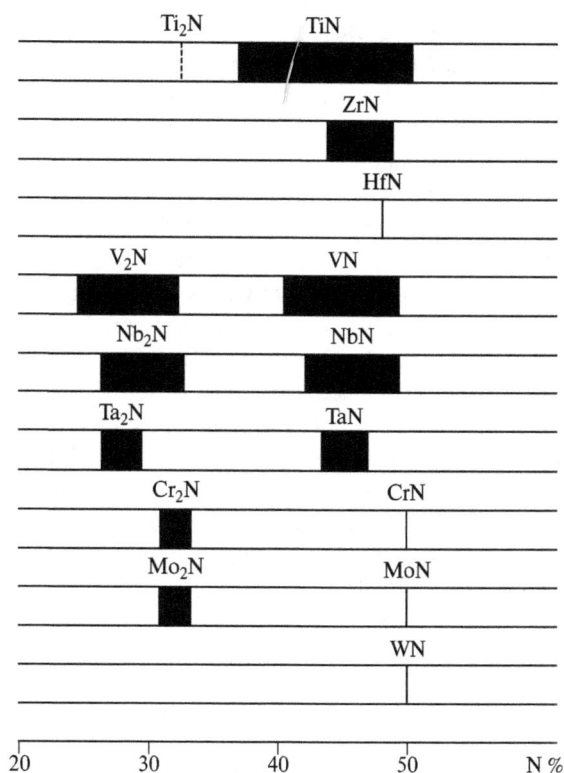

Fig. 29. The homogeneity regions of some nitrides [1]

Despite that in a few cases transition metals form nitrides of a classical composition (e.g. Mn_3N_2). These will not be presented in a separate chapter, but among nitrides called Hägg nitrides, in the order in which the elements forming them are placed in the Mendeleev table. Nitrides which are not interstitial compounds can also be found among some of the actinide species with nitrogen (e.g. PaN_2 or UN_2). Their properties and methods of preparation are presented together in the section devoted to interstitial nitrides of intratransition elements.

Data concerning complex nitrides of the salt character are presented when describing the properties of transition element nitrides. Titanium and metals of the vanadium, chromium and manganese subgroups form compounds containing TiN_3^{5-}, $M^V N_4^{7-}$ or $M^{VI} N_5^{9-}$ anions in their structure. The structure of complex lithium-chromium nitrides may be of interest; five nitride ligands are in the direct surrounding of the coordination center such as chromium, molybdenum or tungsten. These salts are, of course, not interstitial compounds but classical ionic species, composed in the solid state from the anions mentioned above and cationic counterions.

It is assumed that light and heavy platinum metals do not form direct compounds with nitrogen. However, complex nitrides are known, in which osmium or iridium is one of the elements in salts of the $Ba_9Os_3N_{10}$ type or interstitial compounds, where part of the metal atoms in the crystalline sublattice is replaced by another metal, e.g. Mn_3PtN. These latter compounds were obtained in a considerably large number, mainly as derivatives of the interstitial nitride Mn_4N, where a quarter of the manganese is replaced by atoms of another metal.

During the last fifty years a large number of three-component species, besides the interstitial nitrides, were obtained that do not fulfill the rules described earlier; hence they are called non-Hägg nitrides or by the name of the Novotny phases [23].

Nitride compounds of the dsp elements will be presented according to the order of their placing in the periodic table, i.e. in groups (scandium, yttrium, lanthanum, titanium, zirconium, hafnium, etc.). Nitrides of fdsp elements are presented in a separate chapter devoted to lanthanides and actinides. The order of presenting the nitrides is maintained also in this part of the monograph, similarly as in the section concerning the nitride compounds of the main family elements. First of all, the nitride compounds of a described element are presented in the morphological classification expression, if possible and useful. The group that the respective compounds belong to will be marked (interstitial, Novotny phases, ionic). The properties will be presented further as well as the preparation methods of each of the nitride compounds of a given element.

The different electronic structure of transition or intratransition elements compared to that of the main family of elements causes another classification coordinate system to be more convenient to present the structures and transformations joining the species of the dsp and fdsp elements. On the abscissa axis species are ordered according to the changing oxidation degree of the coordination center, not as a function of the e_v value as in the case of the main family elements.

Transformations consisting in a replacement of an electron pair by an oxide ligand (or reciprocally), so often occurring among oxide compounds of sp elements (conjugated red-ac and ox-bas processes), stop to play a dominant role in the case of oxide species of dsp elements. The transition element species can undergo transformations in which the coordination center electron surrounding (red-ox processes) and ligand surrounding (ac-bas processes) undergo changes independently. It should be assumed that similar phenomena may take place among transition metal nitrides. However, until now only a small number of complex nitrides (salts) is known with anions in which a dsp element is the coordination center.

The preparation of a considerable number of new species can be expected in the near future. The possibility of their existence is indicated on the basis of classification predictions when describing the nitrides of elements of consecutive transition families in the periodic system.

The preparation methods of transition element nitrides consist of a few types of reactions carried out at various temperatures for different nitrides. The methods of synthesis will be presented here in a general manner and the short detailed conditions of the procedures in chapters concerning the nitrides of particular elements.

The following routes of obtaining nitrides of dsp and fdsp elements can be distinguished depending on the kind of nitriding reagent used:

1. Nitrogen as the nitriding reagent

    – direct synthesis from elements

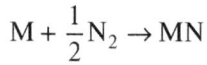

$$M + \frac{1}{2}N_2 \rightarrow MN$$

    – reaction of nitrogen with metal hydride

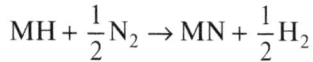

$$MH + \frac{1}{2}N_2 \rightarrow MN + \frac{1}{2}H_2$$

    – reduction of metal oxide or chloride with simultaneous nitriding

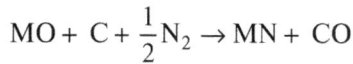

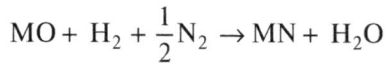

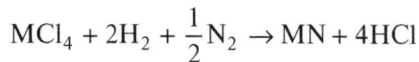

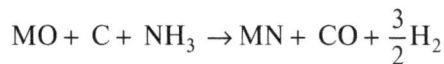

$$MO + C + \frac{1}{2}N_2 \rightarrow MN + CO$$

$$MO + H_2 + \frac{1}{2}N_2 \rightarrow MN + H_2O$$

$$MCl_4 + 2H_2 + \frac{1}{2}N_2 \rightarrow MN + 4HCl$$

2. Ammonia as the nitriding reagent

    – reaction with free metal

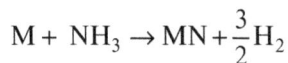

$$M + NH_3 \rightarrow MN + \frac{3}{2}H_2$$

    – reaction with metal hydride

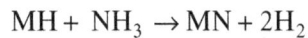

$$MH + NH_3 \rightarrow MN + 2H_2$$

    – reduction of metal oxide or chloride with simultaneous nitriding

$$MO + C + NH_3 \rightarrow MN + CO + \frac{3}{2}H_2$$

$$MCl_4 + \frac{1}{2}H_2 + NH_3 \rightarrow MN + 4HCl$$

5.1. Nitride compounds of the scandium subgroup elements

5.1.1. Scandium nitrides

Scandium forms one compound with nitrogen of the ScN composition. In the crystalline form this nitride has a regular structure of the metal sublattice with nitrogen located in the interstitial sites. Lattice constant a = 4.495Å. It is a chemically stable compound undergoing decomposition in solutions of strong hydroxides with ammonia evolution. Scandium nitride transforms in an oxygen atmosphere at above 600°C into an oxide:

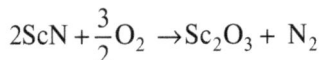

$$2ScN + \frac{3}{2}O_2 \rightarrow Sc_2O_3 + N_2$$

Scandium nitride is obtained in the reaction of scandium oxide with carbon in a nitrogen atmosphere at 1400–2000°C:

$$Sc_2O_3 + 3C + N_2 \rightarrow 2ScN + 3CO$$

A mixture of sodium carbonate with carbon may be applied as additives slightly increasing the yield of the process. Ammonia can be used as the nitriding agent instead of nitrogen at a similar temperature. The synthesis of ScN can also be carried out in fused sodium chloride in a nitrogen or ammonia atmosphere.

Metallic scandium undergoes transformation into nitride in reaction with nitrogen at 1200–1400°C, best under the pressure of several atmospheres. The synthesis of ScN from metal and ammonia is performed at the same temperatures but without pressure increase.

A nitride of a not complete nitrogen content with respect to the theoretical one is obtained in all the methods described; the composition varies in the range $ScN_{0.945-0.970}$. This is characteristic for the majority of interstitial compounds, which was mentioned in the introduction to this chapter.

5.1.2. Yttrium nitrides

Yttrium forms only one interstitial compound with nitrogen – YN. In the solid phase it crystallizes in a regular form of the lattice constant a = 4.887Å. This compound, similarly as scandium nitride, is characterized by a high thermal stability, under the pressure of nitrogen it melts at 2670°C.

Yttrium nitride undergoes slow hydrolysis, when hot, in water and in a water vapor atmosphere. It evolves ammonia during fusion with hydroxides.

It is not possible to obtain the nitride by reducing yttrium oxide with carbon with simultaneous nitriding by means of gaseous nitrogen. The reduction of the oxide does not proceed up to 1500°C, and above that temperature yttrium carbide is formed, which is more stable than nitride. The nitriding of metallic yttrium with gaseous nitrogen proceeds slowly due to the formation of a nitride layer on the metal surface limiting the nitrogen diffusion. The reaction starts at 300°C, but much higher temperatures (up to 1500°C) for a long time (up to a few hours) are required to convert all the yttrium into nitride. Similar difficulties occur during the synthesis of yttrium nitride from the reaction of metallic yttrium with ammonia.

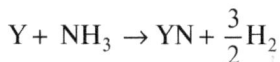

$$2Y + N_2 \rightarrow 2YN$$

$$Y + NH_3 \rightarrow YN + \frac{3}{2}H_2$$

Yttrium nitride of a composition similar to the theoretical one was obtained for the first time from yttrium hydride YH_2 by treating it with gaseous nitrogen at 400–1500°C:

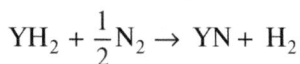

$$YH_2 + \frac{1}{2}N_2 \rightarrow YN + H_2$$

5.1.3. Lanthanum nitrides

Lanthanum forms only one compound with nitrogen of the LaN composition. In the crystalline form this compound has a regular structure of the lattice constant a = 5.294Å. Lanthanum nitride undergoes a slow decomposition process in the presence of water with ammonia evolution:

$$LaN + 3H_2O \rightarrow La(OH)_3 + NH_3$$

and fast decomposition in concentrated aqueous solutions of hydroxides or during fusion with alkali metal hydroxides. It dissolves also in mineral acids with the formation of corresponding lanthanum and ammonium salts. At above 400°C in an oxygen atmosphere or in the air nitride transforms into an oxide with the evolution of free nitrogen.

LaN is obtained from the reactions of metallic lanthanum with nitrogen from 600 to 1700°C, or of lanthanum hydride with nitrogen in a similar temperature range:

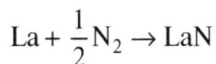

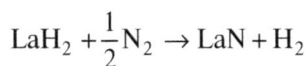

$$La + \frac{1}{2}N_2 \rightarrow LaN$$

$$LaH_2 + \frac{1}{2}N_2 \rightarrow LaN + H_2$$

The application of ammonia as the nitriding agent of lanthanum or its hydride also leads to LaN:

$$La + NH_3 \xrightarrow{1000°C} LaN + \frac{3}{2}H_2$$

$$LaH_2 + NH_3 \rightarrow LaN + \frac{5}{2}H_2$$

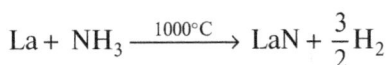

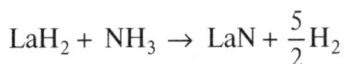

However, pure lanthanum nitride cannot be obtained by a reduction of the oxide with carbon, with simultaneous nitriding with nitrogen or ammonia. Under these conditions a mixture of lanthanum nitride and carbide (with a majority of the latter compound) is obtained.

5.2. Nitride compounds of the titanium subgroup elements

The nitrides of the titanium subgroup elements are characterized by very high mechanical strength and thermal resistance, also under highly corrosive conditions. This concerns interstitial species of the $M^{IV}N$ composition, since only such nitrides are formed by three elements of this subgroup: titanium, zirconium and hafnium. Titanium nitride, Ti_2N, is not a Hägg compound and the complex lithium-titanium nitride, Li_5TiN_3, is known in addition to the above compounds. Li_5TiN_3 – the first salt in which the coordination center (transition element atom) is surrounded by nitride ligands so that the whole forms an anion – starts the series of such compounds of dsp elements.

The properties of titanium nitrides presented below will be supplemented by classification considerations on the possibility of the existence of other, unknown hitherto, mixed nitrides with titanium as the coordination center.

5.2.1. Titanium nitrides

Titanium, as was mentioned above, forms two nitrides: interstitial TiN and Ti_2N not belonging to the Hägg group of compounds. Salts with TiN_3^{5-} anions are also known, i.e. complex nitrides, which can be called nitridetitanates. In Fig. 30. in the $G_{ox} - e_z$ coordinate system (where G_{ox} denotes the oxidation degree, and e_z the number of electrons formally introduced to the coordination center to the formation of bonds by nitride ligands) the titanium nitride species known so far are presented. The cationic species TiN^+ is also marked in the table; it occurs in titanium compounds with nitrogen and halogens: $TiNCl$, $TiNBr$, $TiNI$. The TiN_3^{5-} [21] anion is known in the form of salts with lithium cations [24]. The TiN_2^{2-} anion is presented in a parenthesis between those two species. It has not been obtained until now in any form. TiN_2^{2-} could be formed in two hypothetic reactons:

1. $TiNCl + Li_3N \rightarrow Li_2TiN_2 + LiCl$

$$TiN^+ + N^{3-} \rightarrow TiN_2^{2-}$$

98

2. $Li_5TiN_3 + TiNCl \rightarrow 2Li_2TiN_2 + LiCl$

$$TiN_3^{5-} + TiN^+ \rightarrow 2TiN_2^{2-}$$

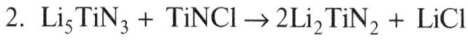

The numbers of the reactions correspond to those of the transformations in Fig. 30. The first one is a simple anionization process and the second one acidic – basic synproportionation. Titanium nitride TiN occurs in the regular form of the crystalline variant of the lattice constant a = 4.240Å. This phase is characterized by the largest region of homogeneity from among all of the transition element nitrides. It exists in the composition range $TiN_{0.56-1.0}$, but it is possible to obtain the maximum nitrogen content $TiN_{0.98}$ (with common methods used in the industry). The properties of titanium nitride, e.g. the reaction temperature with different reactants, depend, to a certain extent, on its composition. The properties of TiN presented below concern the phase of maximum nitrogen content.

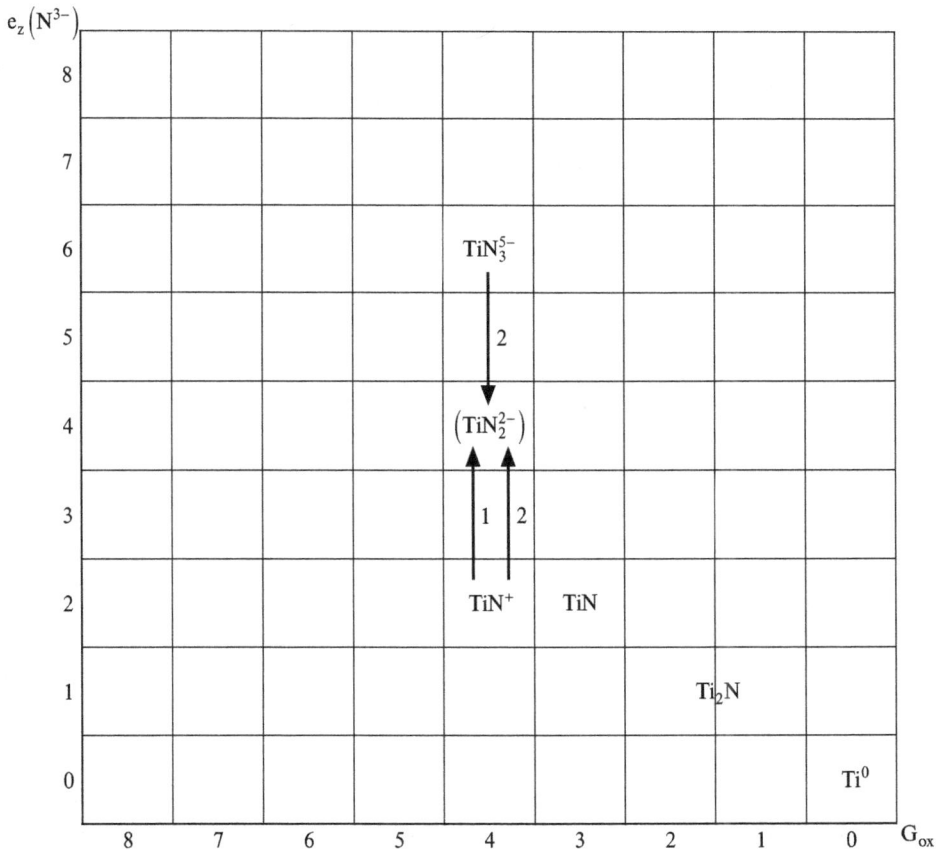

Fig. 30. Nitride compounds of titanium

This compound is stable in the air and in an oxygen atmosphere up to 650°C. Oxidation of the nitride occurs above that temperature according to the reaction:

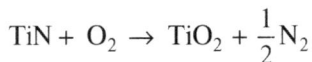

$$TiN + O_2 \rightarrow TiO_2 + \frac{1}{2}N_2$$

At 400°C titanium nitride reacts with chlorine, and above 650°C also with hydrogen chloride:

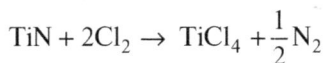

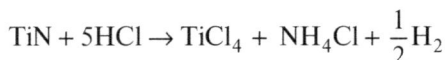

$$TiN + 2Cl_2 \rightarrow TiCl_4 + \frac{1}{2}N_2$$

$$TiN + 5HCl \rightarrow TiCl_4 + NH_4Cl + \frac{1}{2}H_2$$

At 750°C TiN undergoes fast corrosion in the atmosphere of hydrogen sulfide. Above 1500°C titanium nitride transforms under treating with carbon into a carbide; however, an equilibrium is established so as both compounds exist up to ca. 2400°C at increasing content of TiC. Titanium nitride is resistant to strongly reducing agents. Neither hydrogen up to 1000°C nor the majority of fused metals react with TiN.

TiN is obtained by methods used in the synthesis of the majority of nitrides of transition elements. These methods are based on the following reactions (all processes at 500–1200°C):

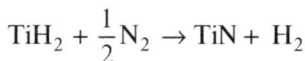

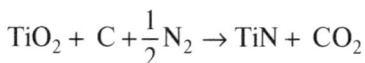

$$Ti + \frac{1}{2}N_2 \rightarrow TiN$$

$$TiH_2 + \frac{1}{2}N_2 \rightarrow TiN + H_2$$

$$TiO_2 + C + \frac{1}{2}N_2 \rightarrow TiN + CO_2$$

Ammonia can be used instead of nitrogen in each of these processes.

The thermal decomposition of the adduct of titanium tetrachloride with ammonia also leads to TiN, and TiNCl is one of the intermediates. Magnesium or calcium hydride can be used as titanium dioxide reducers, with simultaneous nitriding, instead of carbon.

Titanium nitride has a tetragonal structure of lattice constants a = 4.945Å, c = 3.0348Å, untypical for interstitial compounds. Although nitrogen atoms have an octahedral surrounding of titanium atoms, these regular octahedrons are joined via edges into chains, and these form a plain in which every second chain has the vertexes of the octahedrons placed in the plain and the others beyond it. The chemical

100

properties of Ti_2N are described very scarcely in the literature on the subject. It is formed from the reaction of metallic titanium with TiN at above 1000°C; it melts with decomposition at 1450°C. The complex nitride Li_5TiN_3, the anionic part of which (TiN_3^{5-}) is marked in the classification table in Fig. 30., has a regular structure of the lattice constant a = 9.70Å. This compound easily hydrolyzes with ammonia evolution and at elevated temperatures it reacts with oxygen undergoing a transformation into titanate. In reaction with lithium oxide at 900°C it forms a phase of the $Li_5TiN_3 \cdot Li_2O$ composition. Li_5TiN_3 is formed in the reaction of lithium nitride with titanium nitride in a nitrogen atmosphere at 800°C:

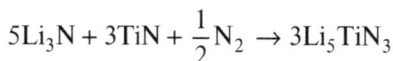

$$5Li_3N + 3TiN + \frac{1}{2}N_2 \rightarrow 3Li_5TiN_3$$

or by nitriding a lithium-titanium alloy:

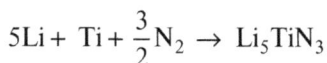

$$5Li + Ti + \frac{3}{2}N_2 \rightarrow Li_5TiN_3$$

Oxynitride compounds of titanium

Phases of variable compositions expressed by the formula TiN_xO_y are formed in the reaction occurring during the synthesis of titanium nitride from titanium dioxide by a reduction with carbon with simultaneous nitriding. Titanium oxynitride is also known. The existence of the crystalline $Li_5TiN_3 \cdot Li_2O$ was mentioned earlier.

The oxide, nitride and oxynitride species of titanium have already been presented in Fig. 31 in the $e_z(O^{2-}) - e_z(N^{3-})$ coordinate system. The hypothetic anions $TiNO^-$, TiN_2O^{4-} and $TiNO_2^{3-}$ are also placed in the classification table. The existence can be predicted on the basis of analogies observed between oxide and nitride species of silicon and titanium. The TiO_2 species corresponds to the SiO_2 species, and the TiO_3^{2-} anion to the SiO_3^{2-} anion. However, among nitride species the analogies are as follows: $TiN^+ - SiN^+$, $TiN_3^{5-} - SiN_3^{5-}$, and among oxynitride species: $Ti_2N_2O - Si_2N_2O$. Li_3TiNO_2, not yet obtained, can be formed in two hypothetic reactions:

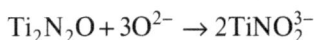

1. $TiO_2 + Li_3N \rightarrow Li_3TiNO_2$

   $TiO_2 + N^{3-} \rightarrow TiNO_2^{3-}$

2. $Ti_2N_2O + 3Li_2O \rightarrow 2Li_3TiNO_2$

   $Ti_2N_2O + 3O^{2-} \rightarrow 2TiNO_2^{3-}$

If among compounds of transition elements the same rules are valid as those observed for oxynitride species of main elements, then the thermal decomposition of those compounds should proceed as follows:

$$3Li_3SiNO_2 \rightarrow 2Li_2SiO_3 + Li_5SiN_3$$

The analogous titanium salt can decompose in a similar manner:

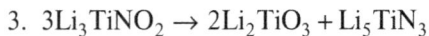

3. $3Li_3TiNO_2 \rightarrow 2Li_2TiO_3 + Li_5TiN_3$

$$3TiNO_2^{3-} \rightarrow 2TiO_3^{2-} + TiN_3^{5-}$$

The numbers of the reactions correspond to those of the transformations marked in Fig. 31. Of course, the whole reasoning requires experimental confirmation.

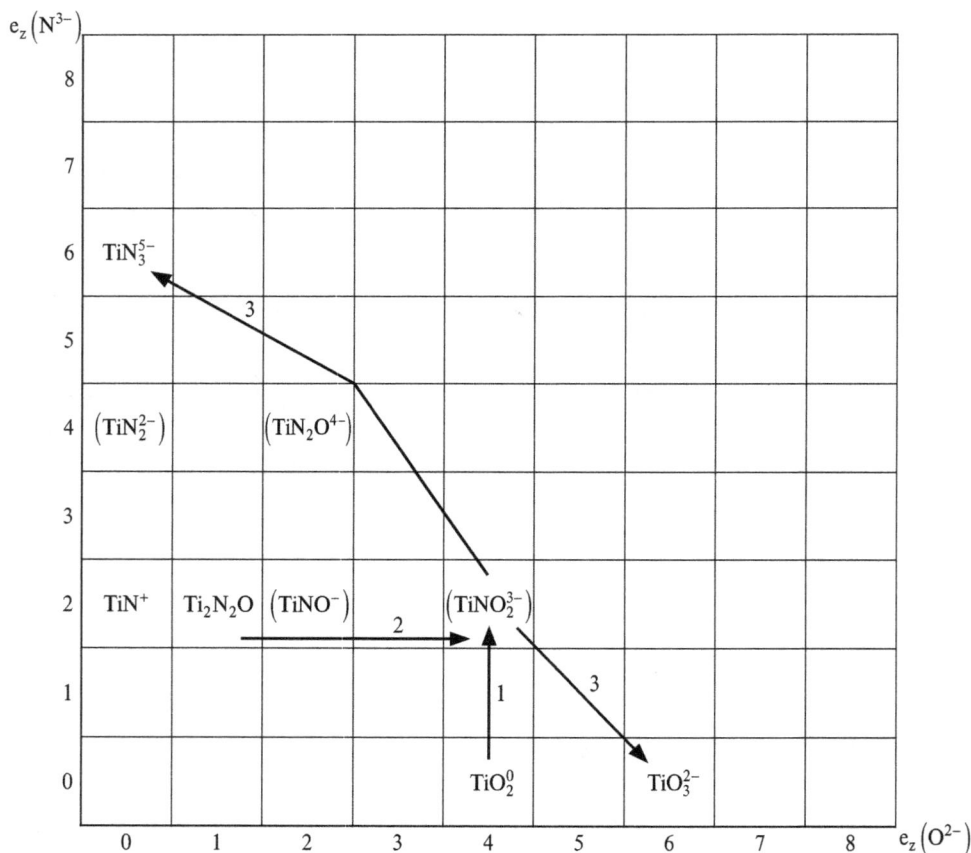

Fig. 31. Oxynitride compounds of titanium

5.2.2. Zirconium nitrides

Two zirconium nitrides are known: interstitial ZrN and probably covalent Zr_3N_4. ZrN has a regular structure of a lattice constant a = 4.577Å and octahedral surrounding of nitrogen atoms by metal atoms, characteristic for Hägg phases. The properties of Zr_3N_4 have not been studied extensively.

Zirconium nitride ZrN melts in a nitrogen atmosphere under normal pressure at 2980°C. It undergoes slow decomposition in vacuum at above 1500°C to free zirconium and nitrogen. The stabilized nitride, i.e. annealed at 1500°C, does not react with oxygen to ca. 1000°C. Above that temperature the following reaction occurs:

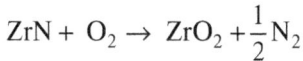

$$ZrN + O_2 \rightarrow ZrO_2 + \frac{1}{2}N_2$$

The above data concern the phase of the ZrN composition. Zirconium compounds with nitrogen are stable in the range $ZrN_{0.67-1.0}$, i.e. in the region of homogeneity only slightly smaller than in the case of titanium nitride. Zirconium nitride reacts above 700°C with chlorine and hydrogen chloride. At above 1100°C ZrN undergoes a reaction with carbon with the formation of the ZrC carbide (more stable than the nitride) and nitrogen evolution. However, zirconium carbide and nitride form solid solutions in a very broad composition range. Zirconium nitride is a compound resistant to the action of mineral acids of any concentration. It also does not decompose under the influence of cold solutions of alkali metal hydroxides. However, it slowly decomposes when hot in concentrated solutions of bases, and also during fusion with alkali metal hydrides. Zirconium nitride is obtained in the following reactions:

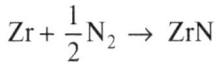

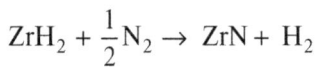

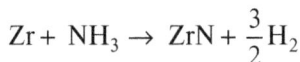

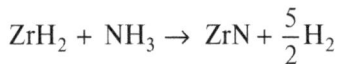

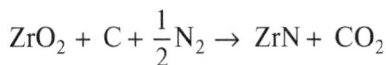

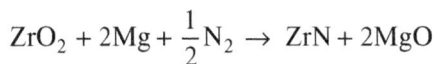

$$Zr + \frac{1}{2}N_2 \rightarrow ZrN$$

$$ZrH_2 + \frac{1}{2}N_2 \rightarrow ZrN + H_2$$

$$Zr + NH_3 \rightarrow ZrN + \frac{3}{2}H_2$$

$$ZrH_2 + NH_3 \rightarrow ZrN + \frac{5}{2}H_2$$

$$ZrO_2 + C + \frac{1}{2}N_2 \rightarrow ZrN + CO_2$$

$$ZrO_2 + 2Mg + \frac{1}{2}N_2 \rightarrow ZrN + 2MgO$$

Zr_3N_4 is formed in the reactions of zirconium halides with ammonia in the 500–1000°C temperature range:

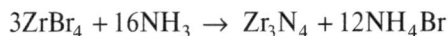

$$3ZrBr_4 + 16NH_3 \rightarrow Zr_3N_4 + 12NH_4Br$$

Zr_3N_4 undergoes decomposition above 1000°C to ZrN with nitrogen evolution:

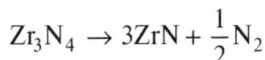

$$Zr_3N_4 \rightarrow 3ZrN + \frac{1}{2}N_2$$

5.2.3. Hafnium nitrides

Only one compound of hafnium and nitrogen has been obtained until now – the interstitial nitride HfN. It has a regular structure of the lattice constant a = 4.526Å and octahedral surrounding of the metal round nitrogen

Hafnium nitride is obtained in the following reactions: the metal with nitrogen or ammonia (at ca. 1200°C), hafnium hydride with nitrogen or ammonia (at ca. 1000°C), hafnium halide with ammonia or a mixture of nitrogen and hydrogen (at 1000°C) and also by the reduction of hafnium dioxide with carbon or magnesium with simultaneous nitriding with nitrogen or ammonia (at 1300°C). The methods applied in the industry for the production of titanium, zirconium and hafnium nitrides are based mainly on the reaction of the metal tetrachloride with a mixture of hydrogen and nitrogen at 700–1000°C:

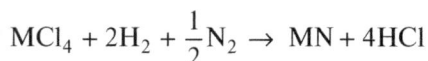

$$MCl_4 + 2H_2 + \frac{1}{2}N_2 \rightarrow MN + 4HCl$$

It can be predicted, similarly as in the paragraph concerning zirconium compounds, that it will be possible to obtain complex nitrides with HfN_3^{5-} anions, M_2N_2O oxynitrides or salts of a mixed oxynitride surrounding of the coordination center. It is assumed that the region of homogeneity of hafnium nitride is very narrow and near the HfN composition. It undergoes slow decomposition in vacuum at above 1300°C to free elements. It melts in a nitrogen atmosphere under normal pressure, with simultaneous decomposition, at ca. 3000°C. It reacts with oxygen above 400°C forming intermediate phases – hafnium oxynitrides of various compositions described by the HfN_xO_y formula. Hafnium nitride reacts with carbon undergoing a transformation into carbide, similarly as zirconium. HfN forms with hafnium carbide solid solutions of unlimited miscibility.

5.3. Nitride compounds of the vanadium subgroup elements

Nitrides of the vanadium subgroup elements, similarly as titanium, zirconium and hafnium nitrides, are characterized by a high mechanical and chemical stability; however, their thermal decomposition points are lower by a few hundred degrees centigrade. This concerns the properties of interstitial nitrides of the MN composition, the most stable among the nitrides of the same element of different compositions. It appears that when a number of nitrides formed by a metal are known, then the most stable is the one where all the accessible interstitial positions are occupied by nitrogen. Thus, VN is the most stable nitride among V_3N, V_2N and VN, NbN among Nb_2N, Nb_3N_4 and NbN, and TaN among Ta_2N, Ta_3N_4 and TaN. This latter nitride and actually one of its types – the ε-TaN phase is not a Hägg compound, since nitrogen is placed beyond the interstitial positions in their classical interpretation.

Vanadium, niobium and tantalum form nitrides with lithium of the $Li_7M^VN_4$ composition besides the nitrides mentioned above. These are the only salts of a purely nitride surrounding of the coordination center known hitherto among the vanadium subgroup elements. It is very probable that it will be possible to obtain other complex compounds, e.g. LiM^VN_2 or $Li_4M^VN_3$ at the maximum oxidation degree of the coordination center equal to 5+. Compounds of this type are known for some time among the complex nitrides of phosphorus, which is presented in the chapter concerning that element ($LiPN_2$, Li_4PN_3 and Li_7PN_4). Predictions on the eventual existence of similar vanadium compounds can be based on some similarities joining oxide compounds of main and transition elements of the same group in the periodic system. The possibility of obtaining vanadium, niobium and tantalum complex nitrides of lower oxidation degree, e.g. $Li_5M^VN_3$, seems to be less probable, but cannot be excluded.

5.3.1. Vanadium nitrides

Vanadium forms three nitrides of the composition: V_3N, V_2N and VN; all of them are interstitial Hägg compounds. They all are characterized by the existence, within a certain range, of an elemental composition. The Li_7VN_4 salt is known, containing VN_4^{7-} anions in its structure.

In Fig. 32. the classification table with the $G_{ox} - e_z$ coordinate system is presented, where the nitride compounds of vanadium are marked. As far unknown species: VN_2^- and VN_3^{4-} are placed (in parenthesis) in the table besides the species described above. Li_7VN_4 was obtained in the reaction of lithium nitride and vanadium nitride in a nitrogen atmosphere at 700°C:

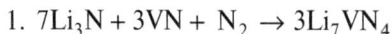

  1. $7Li_3N + 3VN + N_2 \rightarrow 3Li_7VN_4$

Two hypothetic nitride vanadates can probably be obtained in the following reactions:

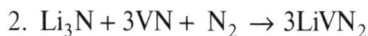

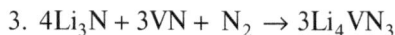

  2. $Li_3N + 3VN + N_2 \rightarrow 3LiVN_2$

  3. $4Li_3N + 3VN + N_2 \rightarrow 3Li_4VN_3$

It is not excluded that acid-basic synproportionation will proceed under the action of nitrogen as the oxidizing agent:

4. $Li_7VN_4 + 6VN + 2N_2 \rightarrow 7LiVN_2$

5. $4Li_7VN_4 + 3VN + N_2 \rightarrow 7Li_4VN_3$

The numbers of the reactions correspond to those in Fig. 32.

The hypothesis of obtaining vanadium nitride (V_3N_5) seems to be probable, more so if Ta_3N_5 is known.

V_2N in the crystalline form has a hexagonal, close-packed structure of lattice constants a = 2.84Å and c = 4.55Å. It is resistant to the action of oxidizing and reducing agents, acids and bases at normal temperature. V_3N is formed during nitriding of metallic vanadium with nitrogen or ammonia at 800–900°C. The structure and properties

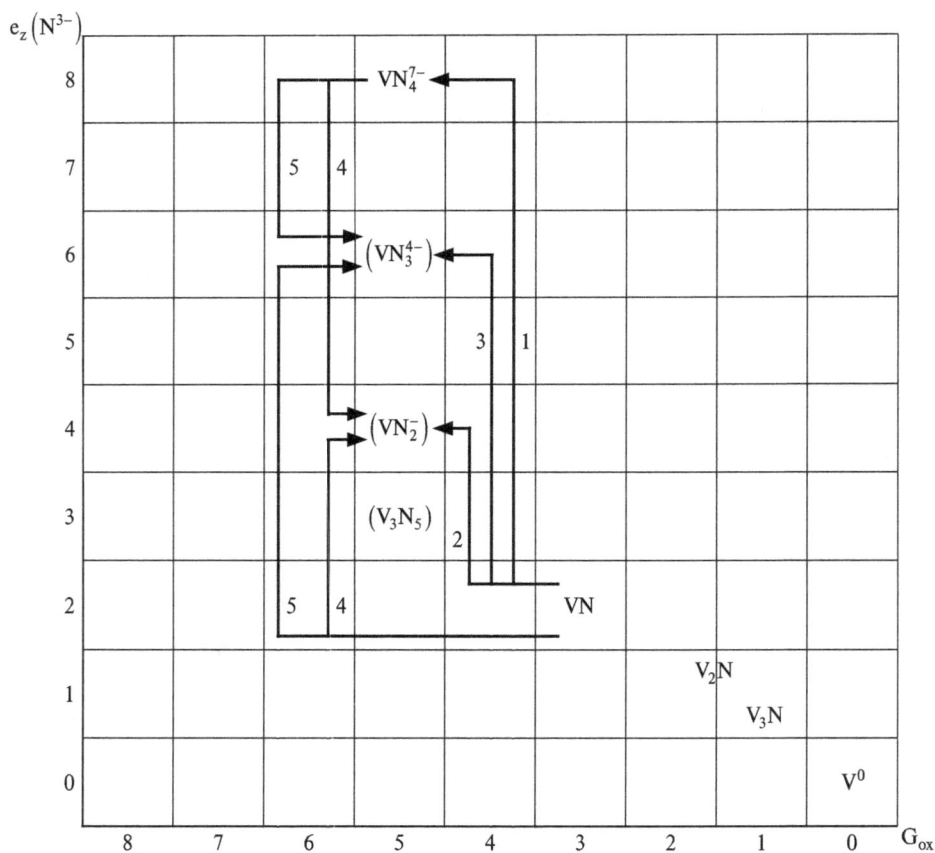

Fig. 32. Nitride compounds of vanadium

of the consecutive vanadium nitride V_2N have not been studied or described. In the crystalline form VN has a regular structure of the lattice constant a = 4.12Å. The phase of this structure exists within a wide range of composition $VN_{0.70-1.0}$. The properties presented below correspond to the nitride of the VN composition. This compounds melts with decomposition at 2050°C (under normal pressure of nitrogen). In vacuum above 1300°C it undergoes slow decomposition to elements.

The VN nitride is obtained both by methods schematically described earlier and used for all the transition element nitrides, as well as for specific ones resulting from the greater variability of the oxidation degrees of vanadium, and thus the greater number of substrates possible to apply. The nitriding of free metal:

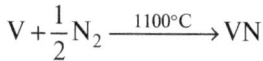

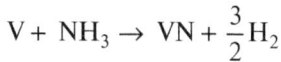

$$V + \frac{1}{2}N_2 \xrightarrow{1100°C} VN$$

$$V + NH_3 \rightarrow VN + \frac{3}{2}H_2$$

or reduction with nitriding:

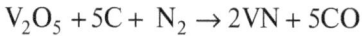

$$V_2O_5 + 5C + N_2 \rightarrow 2VN + 5CO$$

Oxides or other vanadium compounds at lower oxidation degrees are used in the synthesis of VN:

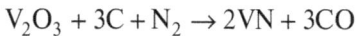

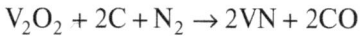

$$V_2O_3 + 3C + N_2 \rightarrow 2VN + 3CO$$

$$V_2O_2 + 2C + N_2 \rightarrow 2VN + 2CO$$

V_2O_3 and V_2O_2 undergo transformations to the nitride also under the action of ammonia or a mixture of nitrogen and hydrogen:

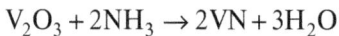

$$V_2O_3 + 2NH_3 \rightarrow 2VN + 3H_2O$$

Similarly VCl_4 with a mixture of nitrogen and hydrogen yields a nitride in the gaseous phase.

Vanadium nitride is also formed in the thermal decomposition of ammonium vanadate at 1100°C in an ammonia and hydrogen atmosphere:

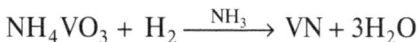

$$NH_4VO_3 + H_2 \xrightarrow{NH_3} VN + 3H_2O$$

or of ammonium hexaflourovanadate

$$(NH_4)_3 VF_6 + 4NH_3 \xrightarrow{\ 600°C\ } VN + 6NH_4F$$

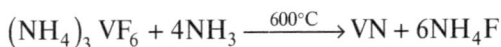

A lithium and lithium nitride alloy can be used as the reducing and nitriding mixture:

$$V_2O_5 + 4Li + 2Li_3N \rightarrow 2VN + 5Li_2O$$

Crystalline phases containing vanadium, nitrogen and oxygen, described by the VN_xO_y formula are known. However, there are no data on the structure and properties of those phases.

5.3.2. Niobium nitrides

Niobium forms three nitrides: Nb_2N, Nb_4N_3 and NbN. A lithium salt, nitride niobate of the formula Li_7NbN_4, is also known [22]. In Fig. 33. the classification table is presented where the known niobium compounds are marked. The hypothetic species which can be formed by niobium are presented in the table (in parenthesis). As far as concerns the crystalline phases which occur in the niobium-nitrogen system – there is a considerable number of them. They correspond to the following compounds:

α – solid solution of nitrogen in niobium in the range from pure niobium to $NbN_{0.025}$;

β – Nb_2N of a hexagonal structure occurring in two forms: richer in nitrogen of lattice constants a = 3.056Å and c = 4.995Å, and poorer in nitrogen of lattice constants a = 3.056Å and c = 4.955Å;

γ – Nb_4N_3 of a tetragonal structure occurring in two forms: richer in nitrogen of lattice constants a = 4.386Å and c = 4.335Å, and poorer in nitrogen of a = 4.385Å and c = 4.310Å;

δ – NbN of a regular structure of the lattice constant a = 4.39Å;

δ' – NbN of a hexagonal structure with lattice constants a = 2.965Å and c = 5.535Å;

ε – NbN of a slight excess of nitrogen (up to the composition $NbN_{1.05}$) and a hexagonal structure of lattice constants a = 2.598Å and c = 11.2728Å.

Nb_2N is a thermally stable compound – it undergoes decomposition in a nitrogen atmosphere at above 2400°C. In oxygen it is stable up to 450°C and at a higher temperature the following reaction takes place:

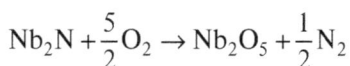

$$Nb_2N + \frac{5}{2}O_2 \rightarrow Nb_2O_5 + \frac{1}{2}N_2$$

Nb_2N does not decompose under the influence of acids solutions nor the hydroxides when cold.

Nb_4N_3 is characterized by a lower thermal stability and at 1500°C it decomposes to:

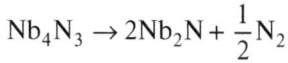

$$Nb_4N_3 \rightarrow 2Nb_2N + \frac{1}{2}N_2$$

In an oxygen atmosphere it transforms into oxide at 500°C:

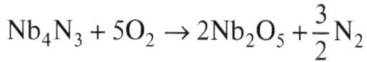

$$Nb_4N_3 + 5O_2 \rightarrow 2Nb_2O_5 + \frac{3}{2}N_2$$

NbN, as was mentioned above, occurs in two forms: δ and ε. At 1230°C the δ-Nb_2N – ε-NbN transformation takes place. The ε-NbN formed is thermally stable up to 2300°C, but above that temperature it decomposes to the elements. In vacuum slow decomposition starts at 1500°C. NbN reacts with oxygen at 500°C according to the reaction:

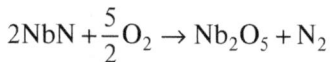

$$2NbN + \frac{5}{2}O_2 \rightarrow Nb_2O_5 + N_2$$

NbN is resistant to the action of hydrogen up to 1000°C and to most of fused metals.

Nitrides of lower nitrogen content are first formed when nitriding metallic niobium with nitrogen or ammonia at 500–1200°C

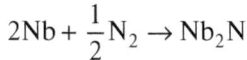

$$2Nb + \frac{1}{2}N_2 \rightarrow Nb_2N$$

Nb_2N can be obtained also in the reaction of NbN with metallic niobium upon long annealing in a nitrogen atmosphere

$$NbN + Nb \rightarrow Nb_2N$$

NbN is formed in the following reactions:

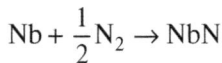

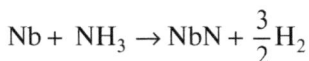

$$Nb + \frac{1}{2}N_2 \rightarrow NbN$$

$$Nb + NH_3 \rightarrow NbN + \frac{3}{2}H_2$$

The reaction with nitrogen is carried out also under a pressure of 300 atm. at 1500°C. It is possible to obtain NbN during the reduction of Nb_2O_5 with simultaneous nitriding:

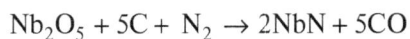

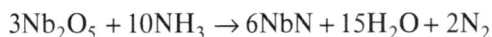

$$Nb_2O_5 + 5C + N_2 \rightarrow 2NbN + 5CO$$

$$Nb_2O_5 + 5H_2 + N_2 \rightarrow 2NbN + 5H_2O$$

$$3Nb_2O_5 + 10NH_3 \rightarrow 6NbN + 15H_2O + 2N_2$$

Phases described as niobium oxynitrides NbN_xO_y are the intermediates of this latter reaction.

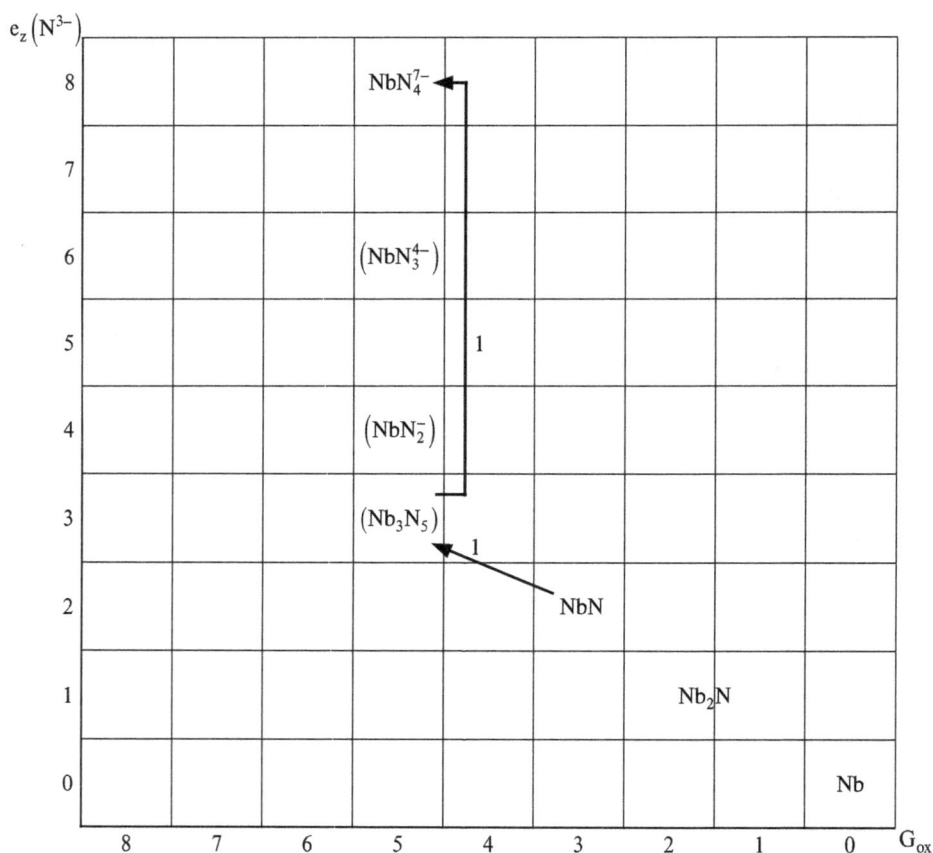

Fig. 33. Nitride compounds of niobium

NbN in reaction with lithium nitride forms a complex nitride Li_7NbN_4 in a nitrogen atmosphere at 700°C.

1. $3NbN + 7Li_3N + N_2 \rightarrow 3Li_7NbN_4$

The direction of the reaction is marked in the classification table in Fig. 33.

5.3.3. Tantalum nitrides

Tantalum forms three two-component nitrides: Ta_2N, TaN and Ta_3N_5. Two of them are interstitial compounds and Ta_3N_5 is a nitride of covalent bonds. The complex lithium-tantalum nitride is also known.

Tantalum nitrides are presented in Fig. 34., in the classification table. The not yet obtained species: TaN_2^- and TaN_3^{4-} are also placed there (in parenthesis). They, as it seems, can exist in the form of salts – complex nitrides. These compounds can be formed in the hypothetic reactions:

1. $Ta_3N_5 + Li_3N \rightarrow 3LiTaN_2$

$Ta_3N_5 + N^{3-} \rightarrow 3TaN_2^-$

2. $Ta_3N_5 + 4Li_3N \rightarrow 3Li_4TaN_3$

$Ta_3N_5 + 4N^{3-} \rightarrow 3TaN_3^{4-}$

Li_7TaN_4, on the other hand, was obtained in the reaction:

3. $3TaN + 7Li_3N + N_2 \rightarrow 3Li_7TaN_4$

4. $Ta_3N_5 + 7Li_3N \rightarrow 3Li_7TaN_4$

The numbers of the reactions correspond to those of the transformations in Fig. 34.

Processes 3 and 4, when carried out under actual conditions, may be a confirmation of the considerations on the possibility of the occurrence of vanadium and niobium nitrides of the $M_3^V N_5$ type as intermediate products in the synthesis of $Li_7M^V N_4$ in the reaction:

$3M^V N + 7Li_3N + N_2 \rightarrow 3Li_7M^V N_4$

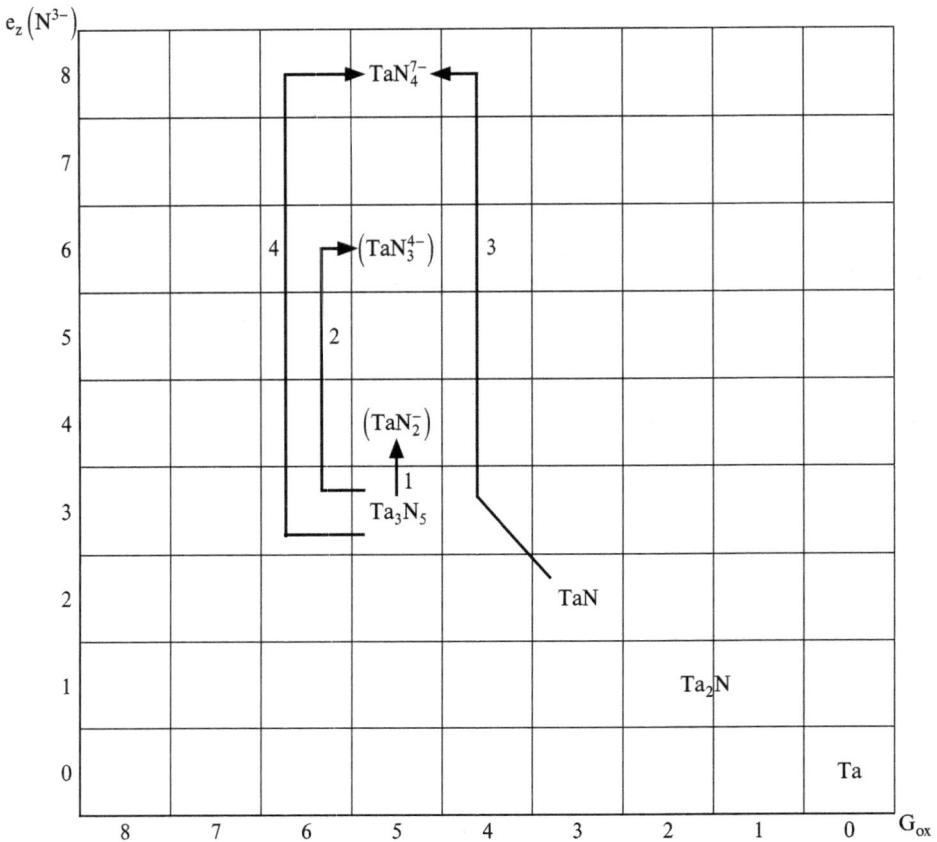

Fig. 34. Nitride compounds of tantalum

Ta_2N has a hexagonal structure in the crystalline form of lattice constants $a = 3.042$Å and $c = 4.905$Å. Phases of this structure exist within the composition range $Ta_{0.41-0.50}$. This compound is resistant to the action of oxygen up to 800°C, and above that temperature it transforms into the oxide:

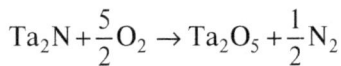

$$Ta_2N + \frac{5}{2}O_2 \rightarrow Ta_2O_5 + \frac{1}{2}N_2$$

Ta_2N is formed by nitriding metallic tantalum with nitrogen or ammonia at 1000°C. It can also be formed in the decomposition of TaN in vacuum at 1200°C:

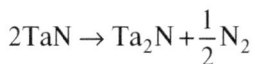

$$2TaN \rightarrow Ta_2N + \frac{1}{2}N_2$$

TaN is formed during nitriding of the metallic tantalum with nitrogen under increased pressure. Nitrides of various compositions are obtained depending on the nitrogen pressure applied in different methods. The other reactions applied for the preparation of interstitial tantalum nitrides are as follows:

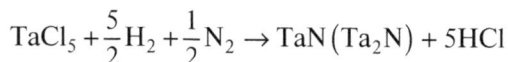

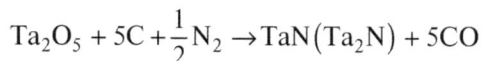

$$TaCl_5 + \frac{5}{2}H_2 + \frac{1}{2}N_2 \rightarrow TaN\left(Ta_2N\right) + 5HCl$$

$$Ta_2O_5 + 5C + \frac{1}{2}N_2 \rightarrow TaN\left(Ta_2N\right) + 5CO$$

Both nitrides (TaN and Ta_2N) are resistant to the action of mineral acids and hydroxide solutions when cold. TaN, similarly to Ta_2N, undergoes oxidation to tantalum pentoxide with oxygen from the air at 800°C with nitrogen evolution.

From the point of view of the classification considerations on nitride compounds of the vanadium subgroup elements, the existence of tantalum nitride at the highest oxidation degree is essential. This compound has a rhombic crystalline structure of lattice constants a = 3.893Å, b = 10.264Å and c = 10.284Å. Ta_3N_5 undergoes decomposition to TaN under vacuum at 500°C and in a nitrogen atmosphere at 900°C:

$$Ta_3N_5 \rightarrow 3TaN + N_2$$

Ta_3N_5 is obtained from the reactions: of tantalum pentoxide with ammonia at 850°C:

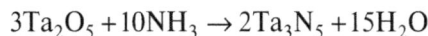

$$3Ta_2O_5 + 10NH_3 \rightarrow 2Ta_3N_5 + 15H_2O$$

tantalum pentachloride with ammonia at 800°C:

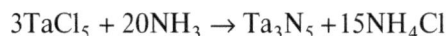

$$3TaCl_5 + 20NH_3 \rightarrow Ta_3N_5 + 15NH_4Cl$$

and also by thermal decomposition of ammonium heptafluorotantalate in an ammonia atmosphere at 800°C:

$$3\left(NH_4\right)_2 TaF_7 + 20NH_3 \rightarrow Ta_3N_5 + 21NH_4F$$

5.4. Nitride compounds of the chromium subgroup elements

Chromium, molybdenum and tungsten nitrides are characterized by a considerably lower thermal and chemical stability in comparison with nitrides of the titanium or vanadium subgroup elements. Therefore, they are applied as construction materials in a very narrow range. Each of the elements of this subgroup forms two two-component compounds with nitrogen: $M_2^{VI}N$ and $M^{VI}N$ and salts – complex nitrides, with lithium of the composition $Li_9M^{VI}N_5$ [24]. These latter compounds are formed in a nitrogen atmosphere from the reaction of a respective nitride with lithium nitride:

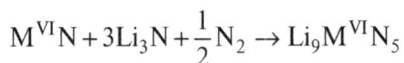

$$M^{VI}N + 3Li_3N + \frac{1}{2}N_2 \rightarrow Li_9M^{VI}N_5$$

In Fig. 35. the classification table is presented, where the nitrides of the subgroup of the sixth group of the periodic system are marked.

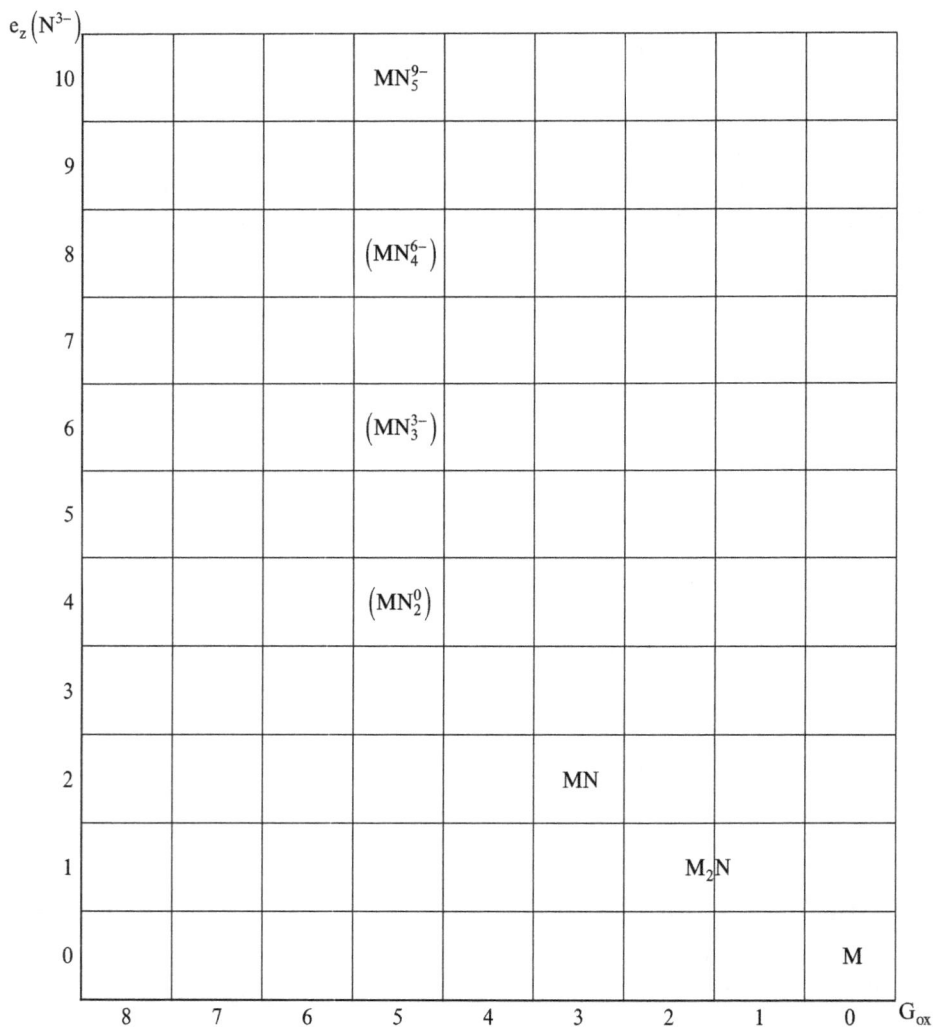

Fig. 35. Nitride compounds of the subgroup elements of the sixth group of the periodic system

The hypothetic compounds, obtaining of which seems to be possible, are also marked. Literature reports do not exclude the possibility of the existence of nitrides $M^{VI}N_2$ or salts with $M^{VI}N_3^{3-}$ or $M^{VI}N_4^{6-}$ anions. The classification relations and analogies between oxide and nitride compounds of the same periods and group elements are the only premises on which these considerations can be based. The mode of reasoning here would be identical as in the case of nitrides of the titanium or vanadium families.

5.4.1. Chromium nitrides

Chromium forms two nitrides: Cr_2N with metallic bonds between the chromium atoms – a typical interstitial compound, and CrN of ionic-covalent chromium-nitrogen bonds and a lack of metallic bonds. A salt, the complex lithium-chromium nitride Li_9CrN_5, is also known; it is formed in the reaction of CrN with lithium nitride and nitrogen at 600°C:

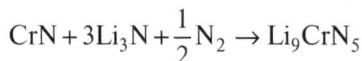

$$CrN + 3Li_3N + \frac{1}{2}N_2 \rightarrow Li_9CrN_5$$

Since placing the nitride species of chromium in the classification table corresponds exactly to the general expression for the subgroup elements of the periodic system sixth group presented in Fig. 35., the table will not be repeated, neither for chromium, nor for molybdenum or tungsten.

Cr_2N has a close packed hexagonal structure in the crystalline form of lattice constants a = 4.79Å and c = 4.47Å. This phase exists in a considerably wide range of compositions $CrN_{0.38-0.50}$. At 1800°C Cr_2N undergoes slow decomposition to elements (in a nitrogen atmosphere). In an oxygen atmosphere it transforms into chromium trioxide at above 600°C:

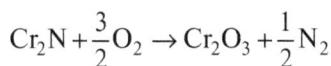

$$Cr_2N + \frac{3}{2}O_2 \rightarrow Cr_2O_3 + \frac{1}{2}N_2$$

Cr_2N is formed during nitriding of metallic chromium with nitrogen or ammonia above 1100°C. The decomposition of CrN under vacuum at above 1100°C leads to Cr_2N:

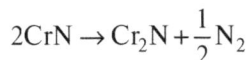

$$2CrN \rightarrow Cr_2N + \frac{1}{2}N_2$$

The Cr_2N phase is resistant to the action of diluted acid solution and hydroxides, but undergoes decomposition under that of concentrated solutions of those substances.

CrN has a regular structure of the lattice constant a = 4.148Å. According to the formulation presented earlier that it is a covalent-ionic compound, it occurs only in the $CrN_{1.0}$ composition, without the region of homogeneity characteristic for interstitial species. The decomposition of CrN leads to Cr_2N. CrN undergoes oxidation in the air above 1200°C – thus it is much more resistant than Cr_2N under these conditions:

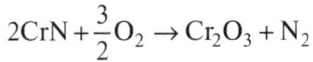

$$2CrN + \frac{3}{2}O_2 \rightarrow Cr_2O_3 + N_2$$

CrN is more stable also in aqueous solutions of acids and hydroxides. CrN can be obtained in the following reactions:

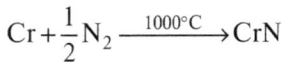

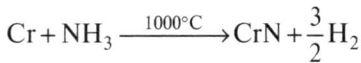

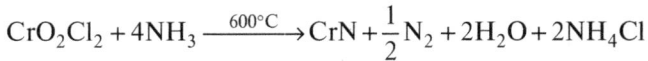

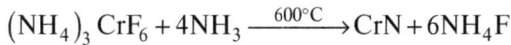

$$Cr + \frac{1}{2}N_2 \xrightarrow{1000°C} CrN$$

$$Cr + NH_3 \xrightarrow{1000°C} CrN + \frac{3}{2}H_2$$

$$CrCl_3 + 4NH_3 \xrightarrow{600°C} CrN + 3NH_4Cl$$

$$CrO_2Cl_2 + 4NH_3 \xrightarrow{600°C} CrN + \frac{1}{2}N_2 + 2H_2O + 2NH_4Cl$$

$$(NH_4)_3 CrF_6 + 4NH_3 \xrightarrow{600°C} CrN + 6NH_4F$$

Li_9CrN_5 easily undergoes hydrolysis in water with ammonia evolutin. It reacts with oxygen at room temperature undergoing transformation into the chromate with nitrogen evolution.

5.4.2. Molybdenum nitrides

Molybdenum forms two two-compounds nitrides: Mo_2N and MoN, both being interstitial compounds. The salt – a complex lithium-molybdenum nitride Li_9MoN_5 is also known. The location of these compounds in the classification table is identical to the general expresion presented in Fig. 35. The salt with the MoN_5^{9-} anion is formed from the following reaction:

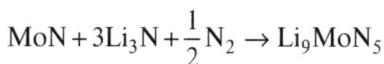

$$MoN + 3Li_3N + \frac{1}{2}N_2 \rightarrow Li_9MoN_5$$

Mo_2N occurs in two crystalline forms: β-MoN, tetragonal of lattice constants a = 4.204Å, b = 8.040Å and c = 1.912Å, and γ-MoN, regular of the lattice constant a = 4.158Å. The γ form is characterized, as most of interstitial compounds, by a region of homogeneity $MoN_{0.40-0.50}$. Mo_2N undergoes decomposition to elements in a nitrogen atmosphere at ca. 900°C, in vacuum, however, the decomposition starts at 700°C. It undergoes oxidation in the air above 500°C according to the reaction:

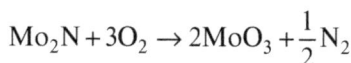

$$Mo_2N + 3O_2 \rightarrow 2MoO_3 + \frac{1}{2}N_2$$

Mo_2N can be obtained in the reaction of metallic molybdenum with nitrogen at above 1100°C and under a pressure of ca. 300 atm. with fast cooling of the product upon completion of nitriding. Similarly, the reaction of molybdenum with ammonia at 400°C during 12h also leads to Mo_2N. However, a mixture of Mo_2N and MoN is obtained in all methods of molybdenum nitride synthesis.

MoN has a hexagonal crystalline structure of lattice constants a = 5.737Å and c = 5.619Å. This compound exists only at a precisely determined composition, with no region of homogeneity. It decomposes to Mo_2N at 850°C under normal pressure of nitrogen:

$$2MoN \rightarrow Mo_2N + \frac{1}{2}N_2$$

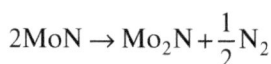

MoN can be obtained in the following reactions:

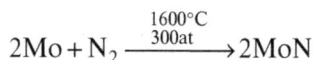

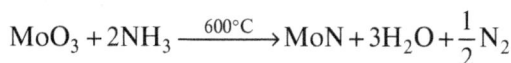

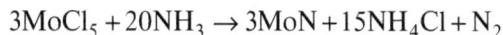

$$2Mo + N_2 \xrightarrow[\text{300at}]{\text{1600°C}} 2MoN$$

$$MoO_3 + 2NH_3 \xrightarrow{\text{600°C}} MoN + 3H_2O + \frac{1}{2}N_2$$

$$3MoCl_5 + 20NH_3 \rightarrow 3MoN + 15NH_4Cl + N_2$$

5.4.3. Tungsten nitrides

Tungsten forms two interstitial nitrides W_2N and WN and complex lithium-tungsten nitride Li_9WN_5. Their placing in the classification table is identical as the chromium and molybdenum nitride species; thus the general expression presented in Fig. 35. also includes tungsten species.

W_2N in the crystalline form has a regular structure of the lattice constant a = 4.118Å. It is stable in a nitrogen atmosphere up to 700°C, above which it undergoes decomposition to the free elements. W_2N undergoes oxidation to WO_3 at elevated temperatures. It decomposes under the influence of concentrated solutions of acids and strong hydroxides. Metallic tungsten at 400°C slowly reacts with nitrogen with the formation of the W_2N phase.

At a slightly elevated temperature (470°C) the nitride WN is the final product. The thermal decomposition of the ammonium tungstate in a nitrogen atmosphere at 550°C also leads to W_2N:

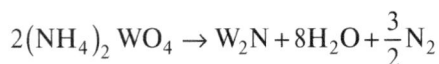

$$2(NH_4)_2 WO_4 \rightarrow W_2N + 8H_2O + \frac{3}{2}N_2$$

WN has a hexagonal structure of lattice constants a = 2.893Å and c = 2.826Å. The WN phase, similary as W_2N, does not show a region of homogeneity; they are stable at a composition corresponding precisely to the stoichiometric one. At 600°C WN undergoes thermal decomposition according to the reaction:

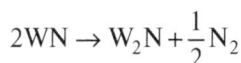

$$2WN \rightarrow W_2N + \frac{1}{2}N_2$$

WN can be obtained in the reaction of metallic tungsten with nitrogen or ammonia at ca. 500 C:

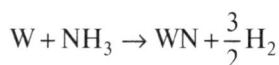

$$2W + N_2 \rightarrow 2WN$$

$$W + NH_3 \rightarrow WN + \frac{3}{2}H_2$$

Other methods of synthesis of tungsten nitrides are applied; they are based on the reduction of WO_3 with metallic lithium or magnesium with simultaneous nitriding. The products of these processes are strongly contaminated by compounds formed in the side reactions.

Li_9WN_5 is slightly more resistant to the action of oxygen or water in comparison to the analogous compounds of chromium and molybdenium.

5.5. Nitride compounds of the manganese subgroup elements

Manganese, technetium and rhenium nitrides are compounds studied scarcely, which has to do with their lower mechanical strength and smaller chemical and thermal resistance with respect to the titanium and vanadium family elements. However, they are good examples and a supplementation to considerations on interstitial nitrides as a separate group of chemical compounds. Manganese forms three nitrides: interstitial Mn_4N and Mn_2N and covalent-ionic Mn_3N_2.

For technetium only one compound was obtained – TcN. Rhenium forms two interstitial nitrides: Re_3N and Re_2N. Beside these two-component compounds the salt Li_7MnN_4 is known among manganese compounds. Mn_4N is a compound from which many phases are derived, with structures formed by replacing one manganese atom by atoms of other, very various metals (the general formula describing these species is Mn_3XN).

5.5.1. Manganese nitrides

Manganese forms three two-elements nitrides: interstitial Mn_4N and Mn_2N and covalent Mn_3N_2. A salt, complex lithium-manganese nitride Li_7MnN_4 was also obtained, in which, contrary to the previously described titanium, vanadium and chromium salts, the central element is not of the maximum oxidation state, in this case Mn^{5+} [22].

In Fig. 36. a classification table is presented, where known manganese species and hypothetic ones (in parenthesis) are marked. Their existence should be expected by the analogy to titanium, vanadium or chromium compounds.

Mn_4N occurs in a regular type of structure of the lattice constant a = 3.686Å. At a normal temperature this compound occurs exclusively at a precise composition, when increasing the temperature; from 400°C a region of homogeneity appears, which at 1000°C reaches $MnN_{0.15-0.25}$.

The nitrides Mn_4N and Mn_2N are not especially thermally stable, in a nitrogen atmosphere they undergo decomposition above 1000°C. Their chemical resistance is also low – they decompose under the action of water when hot, and also of solutions of hydroxides and acids when cold.

The interstitial manganese nitrides are most often obtained by nitriding metallic manganese with nitrogen or ammonia at 600–1200°C or by nitriding the Mn_2Hg_5 amalgam with ammonia. Various compositions of the products are obtained depending on the temperature and pressure of nitrogen. Nitriding amalgam with pure nitrogen leads to Mn_3N_2.

Mn_2N has a hexagonal close–packed structure of lattice constant a = 4.208Å and c = 4.038Å. The region of homogeneity of this phase is quite considerable: $MnN_{0.35-0.50}$.

Mn_3N_2 is formed in the reaction:

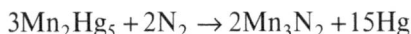

$$3Mn_2Hg_5 + 2N_2 \rightarrow 2Mn_3N_2 + 15Hg$$

A number of different phases also having a regular structure and Mn_3XN compositions is derived from the Mn_4N nitride (X = Cu, Ag, Au, Hg, Pt, Pd, Rh and also Ga or Sn) [3]. Some of the rare nitride compounds of the platinum group elements not forming two-element nitrides are among them.

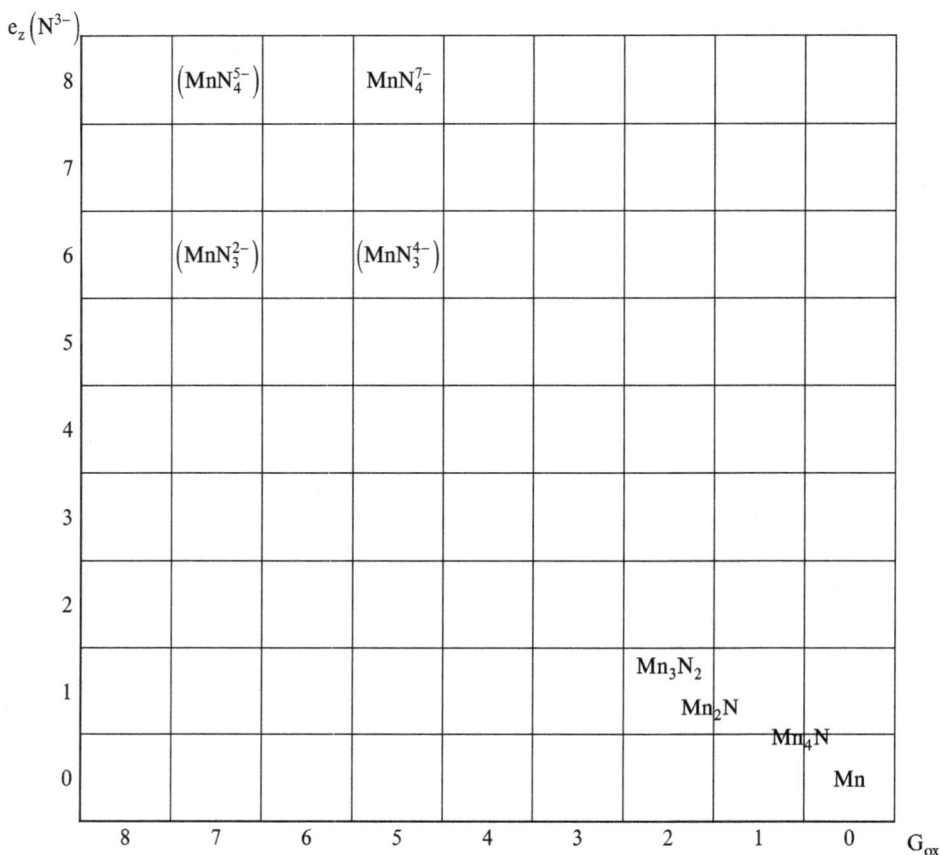

Fig. 36. Nitride compounds of manganese

5.5.2. Technetium nitrides

Technetium forms only one nitride of TcN composition, with a regular structure of the lattice constant $a = 3.93$Å. This compound is formed from the reaction of technetium dioxide with ammonia at 1000°C

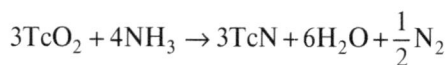

$$3TcO_2 + 4NH_3 \rightarrow 3TcN + 6H_2O + \frac{1}{2}N_2$$

5.5.3. Rhenium nitrides

Two interstitial rhenium nitrides have already been obtained: Re_3N and Re_2N. Re_3N has a regular structure of the lattice constant $a = 3.92$Å.

The thermal stability of both nitrides is low; in vacuum their decomposition starts at 300°C and under normal pressure of nitrogen at 650°C. The chemical properties of both nitrides are similar to those of metallic rhenium. Water does not react with the nitrides in the absence of oxidizing agents. Oxygen oxidizes both phases to species of the oxidation state $+7$.

Another strontium and barium species with $Re_3N_{10}^{18-}$ anions are known besides the two-component nitrides, which are formed from the reaction of strontium or barium nitride with metallic rhenium in a nitrogen atmosphere:

$$3Sr_3N_2 + 3Re + 2N_2 \rightarrow Sr_9Re_3N_{10}$$

Rhenium is formally on the $+4$ oxidation state in this compound. The high electric conductivity of both phases ($Sr_9Re_3N_{10}$ and $Ba_9Re_3N_{10}$) and magnetic properties indicate the occurence of strong metal-metal interactions. Thus it is difficult to include these compounds in the group of salts.

5.6. Nitride compounds of the iron family elements

Iron, cobalt and nickel form the relatively largest number of interstitial nitrides in comparison with the other transition elements: Fe_8N, Fe_4N, Fe_3N, Fe_2N, Co_4N, Co_3N, Co_2N, CoN, Co_3N_2, Ni_4N, Ni_3N and Ni_3N_2. Co_3N_2 and Ni_3N_2 should be treated as ionic-covalent compounds, although this is exclusively of a formal character. There are no data concerning the existence of complex nitrides of the type of salts with the iron family elements as the coordination center.

5.6.1. Iron nitrides

Iron forms four crystalline phases described as interstitial nitrides: Fe_8N, Fe_4N, Fe_3N and Fe_2N.

Fe_8N has a regular structure of the lattice constant $a = 3.65$Å; the region of homogeneity is very broad – $Fe_{0.09-0.128}$

Fe_8N undergoes decomposition at ca. 400°C:

$$8Fe_8N \rightarrow Fe_4N + 4Fe$$

Fe_4N also has a regular structure of a lattice constant $a = 3.795$Å and the region of homogeneity appears at above 400°C.

Fe_3N has a hexagonal close-packed structure of the lattice constants $a = 4.787$Å and $c = 4.418$Å. Fe_3N is formed in the thermal decomposition of Fe_2N in vacuum at 330°C:

$$6Fe_2N \rightarrow 4Fe_3N + N_2$$

Fe_3N decomposes at 370°C according to the reaction:

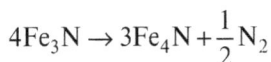

$$4Fe_3N \rightarrow 3Fe_4N + \frac{1}{2}N_2$$

Fe_2N occurs in a rhombic structure of lattice constants a = 2.764Å, b = 4.829Å and c = 4.425Å.

Iron nitrides are characterized by a considerable chemical reactivity. They undergo decomposition under the action of water, acids and hydroxides solutions. Under the influence of chlorine or bromine the nitrides undergo a transformation into the corresponding halides with nitrogen evolution:

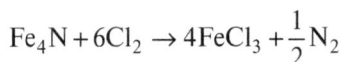

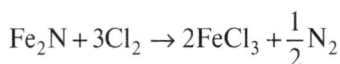

$$Fe_4N + 6Cl_2 \rightarrow 4FeCl_3 + \frac{1}{2}N_2$$

$$Fe_2N + 3Cl_2 \rightarrow 2FeCl_3 + \frac{1}{2}N_2$$

The nitrides react with hydrogen at 200°C.

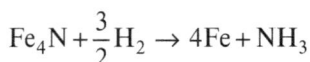

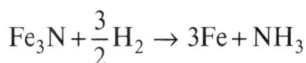

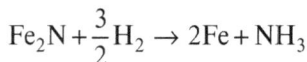

$$Fe_4N + \frac{3}{2}H_2 \rightarrow 4Fe + NH_3$$

$$Fe_3N + \frac{3}{2}H_2 \rightarrow 3Fe + NH_3$$

$$Fe_2N + \frac{3}{2}H_2 \rightarrow 2Fe + NH_3$$

5.6.2. Cobalt nitrides

Cobalt forms four interstitial nitrides: Co_4N, Co_3N, Co_2N and CoN. The chemical properties of these compounds have not been described in detail in the literature on the subject.

Co_3N has a hexagonal structure of lattice constants a = 2.663Å and c = 4.360Å. Co_3N is obtained in the following reactions:

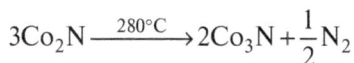

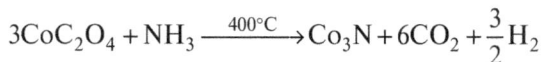

$$3Co_2N \xrightarrow{280°C} 2Co_3N + \frac{1}{2}N_2$$

$$3CoC_2O_4 + NH_3 \xrightarrow{400°C} Co_3N + 6CO_2 + \frac{3}{2}H_2$$

Cobalt triamide decomposes at 60°C with the formation of a Co_3N and Co_2N. mixture. Similar products are obtained when heating a mixture of $Co(CN)_2$ and CoO.

Co_2N has a rhombic structure of the lattice constants a = 2.854Å, b = 4.607Å and c = 4.344Å. It does not exhibit deviations from the composition shown by the formula, similarly as Co_3N. Co_2N is stable up to 290°C, when it decomposes to Co_3N. In solutions of acids both phases slowly dissolve with the formation of cobalt ammonium salts and hydrogen evolution. Co_2N is formed in the reaction of free cobalt with ammonia at 250°C:

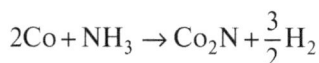

$$2Co + NH_3 \rightarrow Co_2N + \frac{3}{2}H_2$$

5.6.3. Nickel nitrides

Nickel forms three nitrides: two interstitial compounds of the compositions: Ni_4N and Ni_3N, and covalent-ionic Ni_3N_2. Ni_3N has a hexagonal structure of the lattice constants a = 2.670Å and c = 4.307Å. In vacuum it undergoes decomposition to the elements at above 400°C and in a nitrogen atmosphere this process occurs above 800°C.

Ni_3N undergoes transformation into NiO at 750°C under oxygen influence according to the reaction:

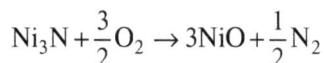

$$Ni_3N + \frac{3}{2}O_2 \rightarrow 3NiO + \frac{1}{2}N_2$$

Nickel nitride slowly dissolves in diluted solutions of acids:

$$2Ni_3N + 7H_2SO_4 \rightarrow 6NiSO_4 + (NH_4)_2 SO_4 + 3H_2$$

Ni_3N is obtained in the follwing reactions:

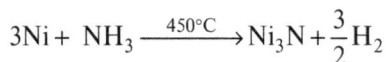

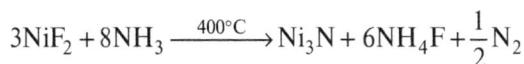

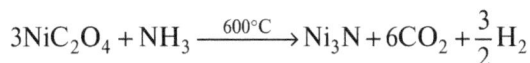

$$3Ni + NH_3 \xrightarrow{\quad 450°C \quad} Ni_3N + \frac{3}{2}H_2$$

$$3NiF_2 + 8NH_3 \xrightarrow{\quad 400°C \quad} Ni_3N + 6NH_4F + \frac{1}{2}N_2$$

$$3NiC_2O_4 + NH_3 \xrightarrow{\quad 600°C \quad} Ni_3N + 6CO_2 + \frac{3}{2}H_2$$

5.7. Nitride compounds of the platinum family elements

There are no confirmed reports on the preparation of two-component nitrides of any elements counted among the light or heavy platinum family elements. However, three-component nitrides were obtained with a few of the elements described. The following compounds of the platinum family elements are included among the complex nitrides. Their structures derive from the Mn_4N phase by replacing one of the four manganese atoms by an atom of another element: Mn_3RhN, Mn_3PdN and Mn_3PtN. Those phases are obtained by heating manganese alloys with rhodium, palladium or platinum in a nitrogen atmosphere. The three-component compound formed in the reaction:

$$Ba_3N_2 + 3Os + 4N_2 \rightarrow Ba_3Os_3N_{10}$$

should be considered as another type of compound. This is an analogous species to the phases of an identical composition formed by rhenium presented earlier.

5.8. Three-component nitrides – Novotny phases

A few dozen of three-component phases of the $M_a M'_b N_c$ composition, called the Novotny phases, are known besides complex nitrides, sometimes called mixed ones. The salts are composed of a cation of an element characterized by very low electronegativity (most often an alkali metal, beryllium, magnesium or alkaline earth metal) and an anion with a coordination like that of an element of a moderate electronegativity value surrounded by nitride ligands. Novotny phases, however, are composed of two metals of moderate electronegativity value and nitrogen. Similarly as in the Hägg compounds, the metal atoms are connected by bonds of a metallic character, formed in a sublattice of the metal. In the part of the octahedral interstitial sites nitrogen is placed. Thus, the octahedron M_6N with metal atoms in its corners and the interstitial site in the center of this octahedron occupied or not is the basic part of structure in the Novotny phases. As in the Hägg compounds the octahedrons with occupied interstitial sites are isolated from each other, in the Novotny phases they are connected by the corners, edges and walls. Complex nitrides, which occur in Novotny phases, can be divided into four groups depending on the formal way of connecting the octahedrons into a structure of the actual compound:

- compounds of the $M_3M'_2N$ type, which can include the following: V_3Zn_2N, V_3Ga_2N, Nb_3Al_2N or four-element $Nb_3(Au_{2/3}Zn_{1/3})_2N_x$ composed of M_6N octahedrons joined by the corners into broken chains.
- compounds of the $M_3M'N$ type, of which the phases: Ti_3InN, Ti_3TlN, Fe_3ZnN, Fe_3GeN, Ni_3AlN etc. are representatives. They are composed of the octahedrons joined by the corners into a closed body, e.g. eight octahedrons joined into a cube involving three corners of each octahedron.

- compounds of the $M_2M'N$ type, to which, e.g. Ti_2AlN, Ti_2GaN, Ti_2InN, Zr_2TlN or Hf_2SnN belong, built of octahedrons joined by edges into a flat slice composed of octahedrons arranged by the angle of 45° of the symmetry axis towards the surface of the slice.
- compounds of the $M_5M'_3N_x$ type, of which $V_5Ge_3N_x$, $Nb_5Ga_3N_x$, $Ta_5Ga_3N_x$ and $Ta_5Al_3N_x$ are representatives, where the M_6N octahedrons are joined via the walls (taking advantage of two walls of each octahedron) into unlimited chains oriented along the c-axis of the symmetry of the whole system [3, 20].

The considerations on the Novotny phases presented above fully concern the second large group of interstitial species – carbides. Thus, like the Hägg compounds, they involve two nearly equivalent groups: nitrides and carbides, among the Novotny phases the compounds with nitrogen and carbon correspond to each other both in composition and relations.

5.9. Nitride compounds of lanthanides

All the lanthanides form nitrides of the MlN composition. In the crystalline form these compounds have a regular structure of a lattice constant varying within a considerably narrow range from 5.164Å for PrN to 4.76Å for LnN. Cerium, praseodymium and neodymium also form nitrides of the MlN_2 composition. These compounds crystallize in a hexagonal type of structure. On the other hand, terbium, dysprosium, holmium, thulium and lutetium, besides MlN, also form compounds of the MlN_x composition, where x is a fraction value between 1 and 2 [1, 2].

These phases are characterized by a regular structure. There are no detailed data on the thermal stability of the nitrides of lanthanides, but generally it is quite high. Cerium nitride, for example, undergoes decomposition in a nitrogen atmosphere involving fusion at 2500°C. In an oxygen atmosphere, lanthanide nitrides undergo transformation into corresponding oxides with nitrogen evolution at relatively low temperatures (200–300°C):

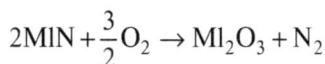

$$2MlN + \frac{3}{2}O_2 \rightarrow Ml_2O_3 + N_2$$

They undergo hydrolysis with ammonia evolution under the influence of water at normal temperatures:

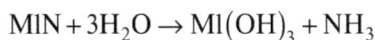

$$MlN + 3H_2O \rightarrow Ml(OH)_3 + NH_3$$

They dissolve in diluted mineral acids and hydroxide solutions.

Lanthanides nitrides are obtained by several methods characteristic for all compounds of this group. These are as follows:

– nitriding of free metal with nitrogen or ammonia at 500–1400°C under normal pressure:

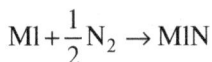

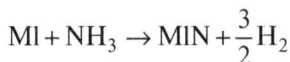

$$Ml + \frac{1}{2}N_2 \rightarrow MlN$$

$$Ml + NH_3 \rightarrow MlN + \frac{3}{2}H_2$$

In some cases nitriding lanthanide metal in the form of an alloy with potassium chloride is applied.

– nitriding lanthanide hydride with nitrogen of ammonia at 500–1400°C
– thermal decomposition of amides

$$Ml(NH_2)_2 \rightarrow MlN + NH_3 + \frac{1}{2}H_2$$

Nitrides of the MlN_2 and MlN_x type are formed in the reaction of metal with nitrogen at 1100–1300°C and under a pressure of at least 30 atm.

5.10. Nitride compounds of actinides

The properties of the nitrides of elements lying beyond actinium in the periodic system are scarcely known. The nitrides of thorium, uranium, neptunium and plutonium were obtained and have been studied very slightly.

Three thorium nitrides are known: ThN, Th_2N_3 and Th_3N_4. ThN has a regular structure of a lattice constant a = 5.21Å. However, Th_3N_4 occurs in two crystalline forms: α – rhombohedral (a = 9.39Å, α = 23.78°) and β – hexagonal (a = 3.830Å, c = 6.200Å). The ThN phase is thermally stable in vacuum up to 2300°C. Th_3N_4 decomposes into ThN at 1750°C. Also Th_2N_3 decomposes with the formation of ThN:

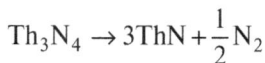

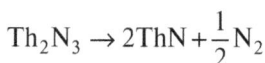

$$Th_3N_4 \rightarrow 3ThN + \frac{1}{2}N_2$$

$$Th_2N_3 \rightarrow 2ThN + \frac{1}{2}N_2$$

All the thorium nitrides slowly react with water and oxygen at room temperature:

$$Th_3N_4 + 3O_2 \rightarrow 3ThO_2 + 2N_2$$

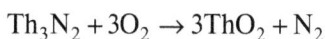

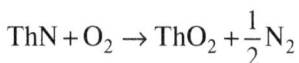

$$Th_3N_2 + 3O_2 \rightarrow 3ThO_2 + N_2$$

$$ThN + O_2 \rightarrow ThO_2 + \frac{1}{2}N_2$$

Th_3N_4 and Th_2N_3 are obtained in reactions of thorium with nitrogen or ammonia in the 350–2400°C temperature range. A mixture of both of the nitrides in various contents is formed depending on the temperature and composition of the gaseous phase.

Th_3N_4 is formed by thermal decomposition of thorium tetraamide at ca. 400°C:

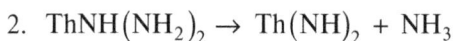

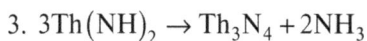

1. $Th(NH_2)_4 \rightarrow ThNH(NH_2)_2 + NH_3$

2. $ThNH(NH_2)_2 \rightarrow Th(NH)_2 + NH_3$

3. $3Th(NH)_2 \rightarrow Th_3N_4 + 2NH_3$

The thermal condensation of thorium amide proceeds identically as the similar process in the case of silicon tetraamide. These transformations can be presented in the classification table (Fig. 37.) using protonless skeletons of species:

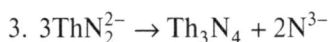

1. $ThN_4^{8-} \rightarrow ThN_3^{5-} + N^{3-}$

2. $ThN_3^{5-} \rightarrow ThN_2^{2-} + N^{3-}$

3. $3ThN_2^{2-} \rightarrow Th_3N_4 + 2N^{3-}$

It can be assumed that by the analogy to the silicon and germanium nitrides, it will be possible to obtain salts with ThN_2^{2-}, ThN_3^{5-} or ThN_4^{8-} anions in reactions of thorium nitride with a strong donor of nitride anions, e.g. Li_3N. The hypothetic reactions can be presented as follows:

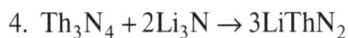

4. $Th_3N_4 + 2Li_3N \rightarrow 3LiThN_2$

5. $Li_2ThN_2 + Li_3N \rightarrow Li_5ThN_3$

6. $Li_5ThN_3 + Li_3N \rightarrow Li_8ThN_4$

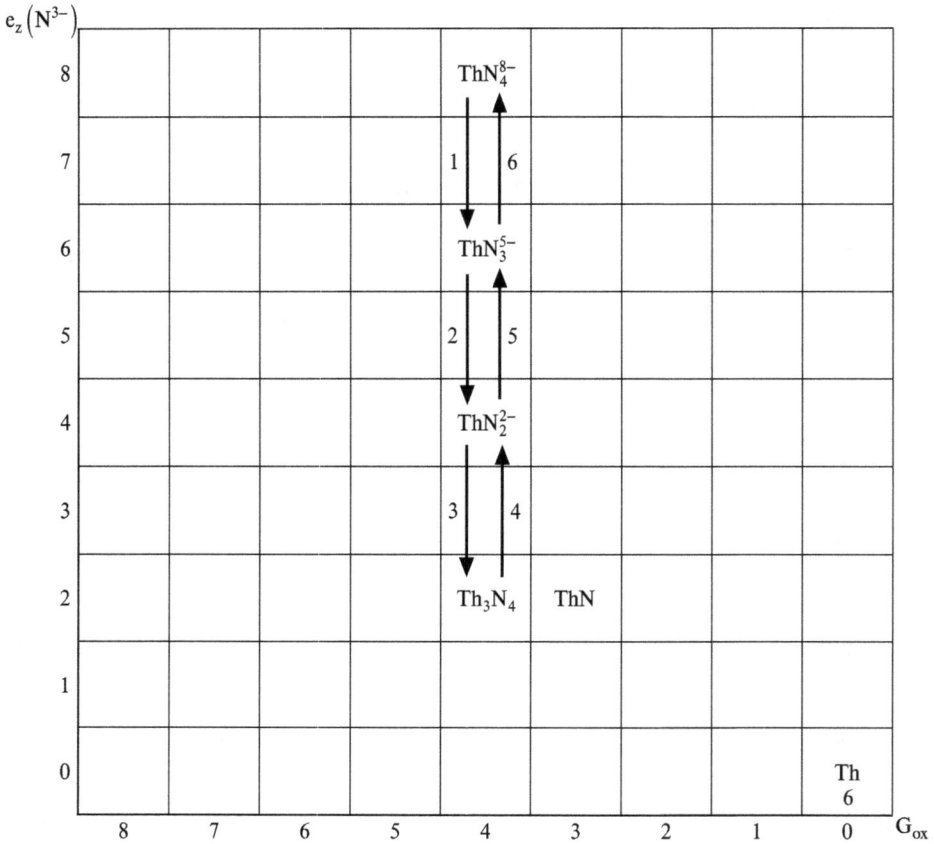

Fig. 37. Nitride compounds of thorium

The directions of transformations (4,5,6) would be contrary to that of the condensation of the amide (1,2,3). The numbers of the reactions correspond to those of the transformations in the classification table.

Uranium forms three compounds with nitrogen: UN, U_2N_3 and UN_2. UN has a regular structure of the lattice constant a = 4.889Å. U_2N_3 occurs in two forms: α and β, α – regular and β – hexagonal. UN_2 has a regular structure of the lattice constant a = 5.21Å.

UN is the most stable phase melting at 2900°C in a nitrogen atmosphere and undergoing decomposition in vacuum above 1800°C. U_2N_3 undergoes transformation to UN above 1300°C:

$$U_2N_3 \rightarrow 2UN + \frac{1}{2}N_2$$

UN_2 decomposes to UN at 1500°C:

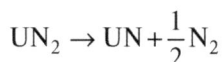

$$UN_2 \rightarrow UN + \frac{1}{2}N_2$$

All uranium nitrides undergo oxidation with oxygen at above 400°C with the formation of U_3O_8 or UO_3 and nitrogen evolution. However, with hydrogen they undergo reduction at a relatively low temperature with the formation of ammonia. They are stable in acids and hydroxides solutions.

Uranium nitrides are obtained in reactions of metal or its hydride with nitrogen or ammonia at 300–1300°C. The phase composition of the products depends on the atmosphere, pressure and temperature.

UN is generally obtained by carrying out the reaction above 1500°C, when nitrogen-rich phases are unstable, and UN_2 is obtained under increased pressure of nitrogen (30 atm.).

Neptunium forms only one nitride NpN – of a regular structure of the lattice constant a = 4.899Å. This phase decomposes in a nitrogen atmosphere at 2700°C. It is resistant to the action of water, but decomposes under the influence of concentrated mineral acids. NpN is obtained in the reaction of neptunium hydride with ammonia at 850°C.

Also plutonium forms only one nitride of a PuN composition. It has a regular structure (a = 4.907Å). It melts with decomposition at 2700°C. This compound is chemically unstable – it reacts with water at room temperature with the formation of plutonium dioxide and evolution of ammonia and hydrogen. However, PuN is resistant to the action of dry oxygen, since this reaction starts at above 500°C:

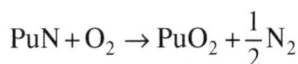

$$PuN + O_2 \rightarrow PuO_2 + \frac{1}{2}N_2$$

Plutonium nitride is obtained in the following reactions (at 1000°C):

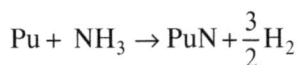

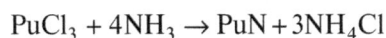

$$Pu + NH_3 \rightarrow PuN + \frac{3}{2}H_2$$

$$PuCl_3 + 4NH_3 \rightarrow PuN + 3NH_4Cl$$

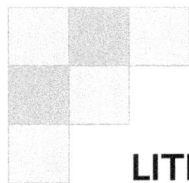

LITERATURE

1. Samsonov G.V., Nitridy, Kiyev, Naukova Dumka,1969
2. Samsonov G.V., Poluceniye y metody analiza nitridov, Kiyev, Naukova Dumka, 1978
3. Toth .L.E., Transition metals carbides and nitrides, NY – London, Academic Press, 1971
4. Gubanov V.A., Ivanovsky A.L., Zhukov V.P., Electronic Structure of Refractory Carbides and Nitrides, 2005, http://assets.cambridge.org/97805214/18850/toc/9780521418850_toc.pdf
5. Podsiadło S. , Azotki , WNT , Warsaw, 1991
6. Gorski A., Chemia, PWN, Warsaw, 1977
7. Kamler G., Weisbrod G., Podsiadło S., Journal of Thermal Analysis and Calorimetry, 2000, **61**, 873
8. Greenwood N.N. , Earnshaw A., Chemistry of the Elements , Elsevier, 2013
9. Greenwood N.N., The Chemistry of Boron, Oxford, Pergamon Press, 1973
10. De Vries R.C., Fleischer J.F., Mater. Res. Bull., 1969, **4**, 433
11. Podsiadło S., J. of Thermal Anal. 1987, **32**, 43, **32**, 445
12. United States Patent 5.110.679, 1992
13. McCauley N.D. , Corbin N.D., J. Am. Ceram. Soc. 1979, **62**, 476
14. Jack K.H., Mat. Res. Bull., 1978, **13**, 1327
15. Wang-Chieh Y., Oxidation Studies of Sialon Ceramics, Stevens Institute of Technology, 2006
16. Brauer G., Handbuch der Preparativen Anorganische Chemie, S. 1 , Stuttgart, Enke, 1981
17. Marchand G. , Ann. Chim., 1985, **10**, 73
18. Edgar J.H., Properties of group III nitrides, INSPEC, Institution of Electrical Engineers, Kansas State University, 1994
19. Morkoç H., Handbook of Nitride Semiconductors and Devices, Vol. 1–3, Wiley-VCH Verlag 2008
20. Hägg G., Z. Phys. Chem., 1931, **Alt B 12**, 33
21. Oyama S.T., The Chemistry of Transition Metal Carbides and Nitrides, Springer, 1996
22. Pierson H.O., Handbook of Refractory Carbides and Nitrides, Elsevier Science, 1996
23. Novotny H., Planseeber. Pulvermet., 1964, **12**, 31
24. Juza R., Z. Anorg. Chem., 1964, **332**, 173, 1961, **302**, 276

www.ingramcontent.com/pod-product-compliance
Lightning Source LLC
Chambersburg PA
CBHW081507200326
41518CB00015B/2418